Memoirs of the American Mathematical Society
Number 349

N. Ghoussoub
and B. Maurey

H_δ-embeddings in Hilbert space and optimization on G_δ-sets

Published by the
AMERICAN MATHEMATICAL SOCIETY
Providence, Rhode Island, USA

July 1986 · Volume 62 · Number 349 (third of 6 numbers)

MEMOIRS of the American Mathematical Society

SUBMISSION. This journal is designed particularly for long research papers (and groups of cognate papers) in pure and applied mathematics. The papers, in general, are longer than those in the TRANSACTIONS of the American Mathematical Society, with which it shares an editorial committee. Mathematical papers intended for publication in the Memoirs should be addressed to one of the editors:

Ordinary differential equations, partial differential equations, and applied mathematics to JOEL A. SMOLLER, Department of Mathematics, University of Michigan, Ann Arbor, MI 48109

Complex and harmonic analysis to LINDA PREISS ROTHSCHILD, Department of Mathematics, University of California at San Diego, La Jolla, CA 92093

Abstract analysis to VAUGHAN F. R. JONES, Department of Mathematics, University of California, Berkeley, CA 94720

Classical analysis to PETER W. JONES, Department of Mathematics, Box 2155 Yale Station, Yale University, New Haven, CT 06520

Algebra, algebraic geometry, and number theory to LANCE W. SMALL, Department of Mathematics, University of California at San Diego, La Jolla, CA 92093

Geometric topology and general topology to ROBERT D. EDWARDS, Department of Mathematics, University of California, Los Angeles, CA 90024

Algebraic topology and differential topology to RALPH COHEN, Department of Mathematics, Stanford University, Stanford, CA 94305

Global analysis and differential geometry to TILLA KLOTZ MILNOR, Department of Mathematics, Hill Center, Rutgers University, New Brunswick, NJ 08903

Probability and statistics to RONALD K. GETOOR, Department of Mathematics, University of California at San Diego, La Jolla, CA 92093

Combinatorics and number theory to RONALD L. GRAHAM, Mathematical Sciences Research Center, AT&T Bell Laboratories, 600 Mountain Avenue, Murray Hill, NJ 07974

Logic, set theory, and general topology to KENNETH KUNEN, Department of Mathematics, University of Wisconsin, Madison, WI 53706

All other communications to the editors should be addressed to the Managing Editor, WILLIAM B. JOHNSON, Department of Mathematics, Texas A&M University, College Station, TX 77843-3368

PREPARATION OF COPY. Memoirs are printed by photo-offset from camera-ready copy prepared by the authors. Prospective authors are encouraged to request a booklet giving detailed instructions regarding reproduction copy. Write to Editorial Office, American Mathematical Society, Box 6248, Providence, RI 02940. For general instructions, see last page of Memoir.

SUBSCRIPTION INFORMATION. The 1986 subscription begins with Number 339 and consists of six mailings, each containing one or more numbers. Subscription prices for 1986 are $214 list, $171 institutional member. A late charge of 10% of the subscription price will be imposed on orders received from nonmembers after January 1 of the subscription year. Subscribers outside the United States and India must pay a postage surcharge of $18; subscribers in India must pay a postage surcharge of $15. Each number may be ordered separately; *please specify number* when ordering an individual number. For prices and titles of recently released numbers, see the New Publications sections of the NOTICES of the American Mathematical Society.

BACK NUMBER INFORMATION. For back issues see the AMS Catalogue of Publications.

Subscriptions and orders for publications of the American Mathematical Society should be addressed to American Mathematical Society, Box 1571, Annex Station, Providence, RI 02901-1571. *All orders must be accompanied by payment.* Other correspondence should be addressed to Box 6248, Providence, RI 02940.

MEMOIRS of the American Mathematical Society (ISSN 0065-9266) is published bimonthly (each volume consisting usually of more than one number) by the American Mathematical Society at 201 Charles Street, Providence, Rhode Island 02904. Second Class postage paid at Providence, Rhode Island 02940. Postmaster: Send address changes to Memoirs of the American Mathematical Society, American Mathematical Society, Box 6248, Providence, RI 02940.

Printed in the United States of America.

The paper used in this journal is acid-free and falls within the guidelines established to ensure permanence and durability.

CONTENTS

Abstract

In this memoir we study the structure of H_δ-subsets of locally convex topological vector spaces. These sets represent the linear analogue of G_δ-sets and share many of the properties enjoyed by convex compact spaces. We examine questions about extremal structures, integral representations and optimization on such sets. Some applications to the geometry of Banach spaces and to non-linear minimization problems are given.

(AMS) (MOS) Subject classification (1980) 46A55, 46B20, 49A27.

Key words: H_δ and G_δ-sets, Martingales, exposed and peak points, linear and Lipschitz perturbations.

Library of Congress Cataloging-in-Publication Data

Ghoussoub, N. (Nassif), 1953–

H_δ-embeddings in Hilbert space and optimization on G_δ-sets.

Memoirs of the American Mathematical Society, ISSN 0065-9266; no. 349)

On t.p. the Greek letter delta is subscript.

"July 1986, volume 62, number 349 (third of 6 numbers)"

Bibliography: p.

1. Locally convex spaces. 2. Embeddings (Mathematics) 3. Martingales (Mathematics) I. Maurey, Bernard. II. Title. III. Series.

QA3.A57 no. 349 510 s [515.7'3] 86-17501

[QA322]

ISBN 0-8218-2350-7

0 INTRODUCTION

Let K be a metrizable convex compact subset of a locally convex topological vector space. It is well known that such a set enjoys the following remarkable properties [6]:

(α) (Krein-Milman): K is the closed convex hull of its extreme points which are actually strongly extreme in K. (i.e. they have a fundamental set of neighborhoods consisting of open slices in K.)

(β) (Choquet): Every point of K is the barycenter of a Radon probability measure supported on the extreme points of K.

(γ) (Bauer): Every concave and lower semi-continuous function on K attains its minimum at an extreme point of K.

In this paper we are mainly concerned with non-compact and sometimes non-convex extensions of the above results. For that we consider a (non-closed) subset C of K such that $K = \overline{\text{conv}}(C)$ and we study the conditions on C that insure the following refinements of α) and β)

(α') K is the closed convex hull of its extreme points that are contained in C.

(β') Every point of C is the barycenter of a Radon probability measure supported on the extreme points of C.

To deal with these problems we need the concept of an H_δ-subset C of K. This is a set such that $K \setminus C = \bigcup_n K_n$ where each K_n is convex and compact. Here are a few examples of such sets:

a) The set of extreme points of K is an H_δ-subset in K. (Proposition I.12).

b) If f is an affine function on K, then the set of points of continuity for f on K is an H_δ-set in K.

Received by the editors March 30, 1985 and, in revised form January 30, 1986.

c) If f is an affine function on K that is also of the first Baire class then any slice determined by f on K (resp. the epigraph of f) is an H_δ-subset of K (resp $K \times \mathbf{R}$) (Proposition II.10).

d) If Y is a Banach space with a norm separable dual **Y***, then any norm closed convex bounded subset C of Y* is an H_δ-set for the w*-topology in its w*-closure K in Y* (Lemma I.1.b).

e) Every separable closed convex bounded set with the Radon-Nikodym property [4] can be "identified" with an H_δ subset of a metrizable compact convex set in some locally convex topological vector space. (Theorem IV.8).

The last statement looks like a "linear analogue" of the well known topological fact that Polish spaces have metrizable compactifications in which they sit as G_δ-sets. We shall see that, in many ways, H_δ-sets are to convex compact sets what G_δ-sets are to compact sets in a non-linear setting.

The reader will immediately notice that the locally convex topology that we always consider is the w*-topology in some dual Banach space Y*. We do it this way for two reasons:

1) No loss of generality occurs for the results in the spirit of α) and β) since any metrizable convex compact subset of a locally convex topological vector space is actually linearly homeomorphic to a weakly compact convex subset of ℓ_2.

2) The second reason is more relevant and has to do with the extension of γ). Indeed, a deep theorem of R. C. James [9] states that a convex bounded subset C of a locally convex t.v.s. E is $\sigma(E,E^*)$ compact if and only if any function in E* attains its maximum on C. Therefore, if we don't assume C to be $\sigma(E,E^*)$-compact, we are led to inquire whether a given continuous linear functional admits "arbitrarily small perturbations" that attain their maximum on K at a point of C. Several results have been already established in this direction: Indeed, if

(C,d) is a complete metric space and if ϕ is a bounded below lower semi-continuous function on C, a theorem of Ekeland [12] gives then that ϕ has an arbitrarily small d-Lipschitz perturbation h such that $\phi + h$ attains its minimum on C. A forerunner of this result is a version of the celebrated Bishop-Phelps theorem due to Brondstëd and Rockafellar [5]: If C is a closed convex bounded subset of a Banach space X and if ϕ is also convex then the perturbation can be taken to be linear and continuous. A remarkable result of Bourgain [1] and Stegall [28] asserts that if C has the Radon-Nikodym property then the same conclusion holds even when ϕ is not assumed to be convex. In this paper, we deal with the following type of problems:

Suppose C is now a subset of a dual space Y*, when can one find arbitrarily small perturbations h that are linear and w*-continuous (i.e. belong to Y as opposed to Y**) (or norm-Lipschitz and w*-continuous), (or C^1-functions) such that $\phi + h$ attains its mininum on C?

The relevance of this bitopological setting is due to the fact that the perturbations can be required to be Lipschitz and small for the strongest metric (the norm) while being continuous for the w*-topology. This naturally explains why the norm is involved in the definition of the sets on which such "perturbed maximum principles" hold (i.e. the concept of strong H_δ-sets defined below).

We start first with the needed definitions.

<u>Definition 0.1</u>: Let Y be a Banach space and let C and D be two subsets of Y* such that $C \subseteq D$. We shall say that:

(i) C is a <u>w*-G_δ</u> (resp <u>w*-H_δ</u>) in D if $D \setminus C = \bigcup_n (K_n \cap D)$ where each K_n is w*-compact (resp w*-compact and convex) in Y*.

(ii) C is a strong w^*-G_δ (resp a strong w^*-H_δ) in D if the K_n's can be chosen to be a strictly positive $\| \ \|$-distance away from C.

Note that if D is w^*-compact (resp w^*-compact and convex) then C is a w^*-G_δ (resp w^*-H_δ) in D if and only if $D \setminus C = \bigcup_n K_n$ where each K_n is w^*-compact (resp w^*-compact and convex). If Y is separable, then w^*-compact subsets of Y^* are w^*-metrizable, w^*-H_δ subsets of Y^*. It follows that a bounded (resp a bounded and convex) set C is a w^*-G_δ (resp a w^*-H_δ) in its w^*-closure if and only if it is a w^*-G_δ (resp a w^*-H_δ) in Y^*.

Before summarizing the main results, we recall the following notions.

Throughout this paper, the notation $\bar{C}^*$ will always mean the w^*-closure of C in Y^*.

Definitions 0.2: Let C be a bounded subset of a dual space Y^* and let x be an element of C. Then x is said to be

(i) An extreme point of C iff $x \notin \text{conv}(C \setminus \{x\})$

(ii) A w^*-strongly extreme point of C iff $x \notin \overline{\text{conv}}^*(C \setminus V)$ for each w^*-neighbourhood V of x.

(iii) A w^*-denting point of C iff $x \notin \overline{\text{conv}}^*(C \setminus B(x,\varepsilon))$ for all $\varepsilon > 0$.

(iv) A w^*-exposed (resp strongly w^*-exposed) point in C iff there exists y in Y such that

(a) $y(x) = \sup y(C)$

(b) If (x_n) is a maximizing sequence for y (i.e. $\lim_n y(x_n) = \sup y(C)$)

then (x_n) w^*-converges (resp norm-converges) to x.

Any function (not necessarily in Y) that verifies a) and b) is said to be exposing (resp strongly exposing) C from above at x. We define in a similar fashion the functions that expose C from below.

(v) A strong w^*-peak point in C iff there exists a w^*-continuous and norm-Lipschitz function on $\bar{C}^*$ that strongly exposes C from above at x.

Recall that a w*-slice of C is a set of the form $S(C,y,\alpha) = \{x \in C; y(x) \geq \sup y(C) - \alpha\}$ where y is in Y and α is a strictly positive real number. Note that by the Hahn-Banach theorem, a point x is w*-strongly extreme in a convex bounded set C if and only if it has a fundamental set of w*-neighborhoods in C consisting of w*-open slices. Moreover such points are exactly the extreme points of the w*-closure of C that are contained in C. Note also that w*-exposed points are necessarily w*-strongly extreme and that strongly w*-exposed points are strong w*-peaks which in turn are points of weak* to norm continuity for the set C. It is also well known that all these notions are distinct ([4], [10]).

In section I we study the extremal structure of convex w*-H_δ sets. Assume that C is a bounded convex subset of a dual space Y* and that it is contained in a w*-metrizable w*-compact convex set D. We first show that:

1) C is a line-closed w*-H_δ in D if and only if C is a w*-G_δ set which is also <u>w*-martingale compact</u>: that is every finitely valued martingale in C w*-converges to a w*-measurable random variable which is almost surely valued in C.

These properties give w*-H_δ sets an extremal structure very much like w*-compact convex sets. We get for instance:

2) If C is a line-closed w*-H_δ set then it is contained in the weak*-closed convex hull of its w*-strongly extreme points.

Moreover, if C is also assumed to be w*-measure convex then the following Choquet-type integral representation holds:

3) Every point in C is the barycenter of a w*-Radon probability measure supported on the extreme points of C.

On the other hand, w*-H_δ sets verify the following "convex analogue" of the Baire Category theorem:

4) If (C_n) is a sequence of convex line-closed w*-dense w*-H_δ sets in D, then $D = \overline{\text{conv}}^*(\bigcap_n E_n)$ where each E_n is the set of extreme points of C_n.

The analogy with w*-compact convex sets also appears in the following extension of a Theorem of Stegall [26]:

5) A bounded convex and line-closed w*-H_δ set in D is norm separable and w*-dentable whenever it contains no δ-tree for any $\delta > 0$.

6) Conversely, any separable closed convex subset of D is a w*-H_δ set in D whenever it is w*-dentable.

In section II, we deal with the structure of strong w*-H_δ sets. This class of sets seems to give the right setting for optimization problems. We now assume that C is a bounded (not necessarily convex) subset of Y* such that C is a strong w*-H_δ in D = $\overline{\text{conv}}^*(C)$. We first show that:

7) If D is w*-metrizable (resp if C is norm-separable) then D is the w*-closed convex hull of its w*-exposed (resp strongly w*-exposed) points that are contained in C.

We then let $\mathcal{H}(C)$ be the class of extended real-valued functions on C whose epigraph is a strong w*-H_δ in C × **R**, and we obtain:

8) If D is w*-metrizable (resp if C is norm-separable) then for any bounded below function ϕ in $\mathcal{H}(C)$ the set $\{y \in Y; \phi + y$ exposes (resp strongly exposes) C from below$\}$ is a dense G_δ in Y.

Several interesting facts follow from this result when one notes the following:

(a) Any w*-lower semi-continuous function belongs to $\mathcal{H}(C)$ if D is w*-metrizable and if C is norm separable then any norm-lower semi-continuous function is also in $\mathcal{H}(C)$.

(b) If ϕ is concave and w*-lower semi-continuous, then 8) implies that the set $\{y \in Y; \phi + y$ exposes C from below at an extreme point of C$\}$ is a dense G_δ in Y. This can be seen as a non-convex compact analogue of Bauer's maximum principle [6].

(c) Any w*-Baire(1) function in Y** belongs to $\mathcal{H}(C)$. Hence (8) applied to such functions gives a new result even when C is assumed to be w*-compact and convex.

As a corollary of (c) we get the following extension of a theorem of Haydon ([4] Theorem 7.4.9e).

9) If ℓ_1 does not embed in Y then every convex strong w*-H_δ set in D is the norm closed convex hull of its exposed points.

We conclude section II by giving the following converse to 8).

10) Assume that C is a w*-G_δ in D. Then C is a strong w*-H_δ in D if and only if for any convex bounded below and w*-lower semi-continuous function ϕ on C the set $\{y \in Y; \phi + y$ attains its minimum on C$\}$ contains a dense G_δ in Y.

<u>Definition 0.3</u>: Let C and D be two subsets of a Banach space X such that $C \subseteq D$. We shall say that <u>C is a norm H_δ in D</u> if there exists a sequence of closed convex bounded subsets (F_n) in X such that $D \setminus C = \bigcup_n F_n \cap D$. If the F_n's can be chosen to be a strictly positive distance away from C, we then say that C is a <u>strong norm H_δ-set in D</u>.

In section III, we exhibit few cases where a norm H_δ (resp a w*-H_δ) set is necessarily a strong norm H_δ (resp a strong w*-H_δ) set. The main result is the following:

11) Let C be a bounded subset of a dual Banach space Y*. Assume either that C is norm separable or that $D = \bar{C}^*$ is w*-metrizable. Then C is a strong w*-H_δ set in D if and only if it is a w*-H_δ and verifies the <u>*-Brondsted-Rockafellar property</u>: that is for any bounded convex w*-l.s.c. function ϕ on C, the set $F(\phi,C) = \{y \in Y;\ \phi + y$ attains its minimum on C$\}$ is dense in Y.

Note that we do not assume that the set $F(\phi,C)$ is a G_δ in Y. Hence any closed convex bounded subset of a Banach space X has the *-Brondsted Rockafellar property when considered as a subset of X**. As applications of 11) we get the following:

(a) Any separable closed convex bounded subset C of a Banach space X is a strong norm H_δ in X whenever it is a norm H_δ-set in X. Moreover, C will be a strong w*-H_δ in its w*-closure D in X** whenever it is a w*-H_δ in D.

(b) If C is a closed convex bounded subset of a dual space Y* such that ℓ_1 does not embed in Y and if C is either norm separable or with a w*-metrizable w*-closure D, then C is a norm H_δ in D if and only if C is a strong w*-H_δ in D.

By using different methods we can also prove the following:

12) If C is the ball of a separable closed subspace X of the dual of a separable Banach space Y* such that C is w*-dense in B_{Y*}, then C is a strong w*-G_δ (resp a strong w*-H_δ) in B_{Y*} whenever it is a w*-G_δ (resp a w*-H_δ) in B_{Y*}.

We shall see in section VI that the notions of w*-H_δ and strong w*-H_δ sets are essentially different even for separable closed convex bounded sets.

At the beginning of section IV, we give a first representation of separable sets with the <u>point of weak to norm continuity property</u> (P.C.P) (resp the <u>Radon-Nikodym property</u> (R.N.P) as w*-G_δ (resp w*-H_δ) sets in some dual space Y*.) For (P.C.P.) sets we only need the definition: That is a set has (P.C.P.) whenever any non-empty subset of C has weak neighborhoods of arbitrarily small norm-diameter. However, for (R.N.P) sets, it is the "vector measure" definition that is used in section I. Hence, we need the well known fact that (R.N.P.) closed convex sets, are <u>dentable</u> ([4] Theorem 2.3.6): that is all their non-empty closed convex subsets have slices of arbitrarily small diameter. The first representation gives then the following:

13) If C is a separable closed bounded P.C.P. (resp R.N.P.) subset of a Banach space X and if X has a countable norming set, then there exists a separable subspace Y of X* such that X is isometric to a subspace of Y* in such a way that C is a w*-G_δ (resp a w*-H_δ) set in Y*.

Actually, the space Y is a separable subspace of X* containing the functions that determined weak neighborhood (resp slices) of C of arbitrarily small diameter in a somewhat "random splitting" (resp "slicing") of C. The above representation and a classical theorem of Baire (resp Theorem 5 mentioned above) give that Y is "big enough" to give every norm closed subset of C points of weak* to norm continuity (resp w*-denting points). However the example in Section VI of a separable bounded closed convex set which is a w*-H_δ set but not a strong w*-H_δ set, shows that Y may not be large enough to "expose" the set C unless we are in the special cases mentioned in section III (i.e. C is the ball of X or ℓ_1 does not embed in X*). To represent C as a strong w*-H_δ set we need to enlarge Y by adding to it a suitable countable set of functionals obtained by a repeated use of the Bishop-Phelps theorem. This is done at the end of section IV. Several old and new results about R.N.P. sets follow from this representation and the results of section III.

14) A separable Banach space has R.N.P. if and only if it has the <u>asymptotic-norming property</u>; a notion introduced by James-Ho [18].

15) If Y is a separable Banach space not containing ℓ_1, then any closed convex bounded subset of Y* which contains no δ-tree for any $\delta > 0$, is contained in the w*-closed convex hull of its strongly w*-exposed points: This extends a result of Bourgain [1].

16) If C is a separable closed convex R.N.P. set in a Banach space X then there exists a separable subspace Y_∞ of X* such that for any closed subspace Z of X* containing Y_∞, the set $\{y \in Z;\ y \text{ strongly exposes } C\}$ is a dense G_δ in Z.

In section V we deal with subsets of Banach lattices. We first show that if K is a metrizable compact Hausdorff space and if C is a separable closed convex subset of $\mathcal{M}(K)$ then

17) C is a w*-H_δ in $\mathcal{M}(K)$ if and only if it is a w*-G_δ.

This result is used to show that

18) For closed convex bounded subsets of Banach lattices not containing c_0, the (P.C.) and the (R.N.) properties are equivalent.

We then give a simple proof of a result of Talagrand [29] stating that a separable Banach lattice with R.N.P. is isometric to a dual Banach lattice.

In section VI we study the structure of strong w^*-G_δ sets by associating to them suitable strong w^*-H_δ sets. We obtain the following results:

19) If C is a separable bounded w^*-G_δ set in its w^*-closure D in Y^* and if D is w^*-metrizable, then C is a strong w^*-G_δ set in D if and only if it verifies the <u>*-Ekeland property</u>: Every bounded below, lower semi-continuous function ϕ on C has an arbitrarily small perturbation h which is w^*-continuous and norm-Lipschitz such that $\phi + h$ exposes C from below.

20) Under the above hypothesis, the set of strong w^*-peak points of C is w^*-dense in C.

21) If X is a separable Banach space with the (P.C.) property, then for any closed bounded subset C of X and any bounded below lower semi-continuous function ϕ on C, there exists an arbitrarily small perturbation h which is weakly continuous and norm Lipschitz such that $\phi + h$ exposes C from below. Actually h can be chosen to be a C^1-function.

22) There exists a bounded separable w^*-G_δ subset of ℓ_∞ which is not a strong w^*-G_δ, from which we can deduce the existence of a w^*-H_δ set which is not a strong w^*-H_δ.

In section VII, we investigate a procedure that reduces non-linear optimization problems to a linear setting where the above results can be used.

We introduce the following bitopological setting

<u>Definition 0.4</u>: A space K equipped with two metrics Δ_1 and Δ_2 is said to be a <u>bi-metric space</u> if Δ_2 is Δ_1-lower semi-continuous on $K \times K$. A subset C of K will be called a <u>Δ_2-strong Δ_1-G_δ</u> in K if $K \setminus C = \bigcup_n K_n$ where each K_n is Δ_1-closed and Δ_2-$\text{dist}(K_n, C) > 0$.

We show that every bi-metric space (K, Δ_1, Δ_2) such that (K, Δ_1) is compact and (K,Δ_2) is complete can be "embedded" in the dual of a Banach space Y in such a way that (K,Δ_1) "corresponds" to a w*-compact in Y* and Δ_2 is "equivalent" to the norm. The Δ_2-strong Δ_1-G_δ sets in K correspond then to strong w*-G_δ sets in Y*. Moreover Y is the space of Δ_1-continuous and Δ_2-Lipschitz functions on Y. In view of Theorem (19) we get the following "bitopological Ekeland's theorem".

(23) If C is a Δ_2-separable Δ_2-strong Δ_1-G_δ subset of a bi-metric space (K,Δ_1,Δ_2) such that (K,Δ_1) is compact and (K,Δ_2) is complete then for any bounded below Δ_2-lower semi-continuous function ϕ on K and for any $\varepsilon > 0$, there exists a Δ_1-continuous and Δ_2-Lipschitz function h with $\|h\|_{Lip(\Delta_2)} \leq \varepsilon$ and an x_0 in C such that

(i) $\phi + h$ attains its minimum on C at x_0.

(ii) Every minimizing sequence for $\phi + h$, Δ_2-converges to x_0.

Note that if C is a Δ_2-separable Δ_1-G_δ in K, the Baire Category theorem gives that C has a Δ_1-dense set of points of (Δ_1 to Δ_2)-continuity. For strong G_δ-sets, we obtain the following.

(24) Under the hypothesis of (23), C contains a Δ_1-dense set of <u>$(\Delta_1$-$\Delta_2)$ peak points</u>: These are the points x such that there exists a Δ_1-continuous and Δ_2-Lipschitz function ϕ that strongly exposes C at x. Note that $(\Delta_1$-$\Delta_2)$ peak points are necessarily points of Δ_1 to Δ_2 continuity for C.

We show that any separable complete metric space (C,d) is a Δ_2-strong Δ_1-G_δ in some bi-metric space verifying the hypothesis of (23). This allows us to obtain Ekeland's theorem as a particular case of (23).

(25) We finally establish that if $\Delta_2 \geq \Delta_1$ and if C is a Δ_2-strong Δ_1-G_δ in K, then there exists a metric Δ on K such that (K,Δ,Δ_1) is a bi-metric space, $\Delta_2 \geq \Delta$ and Δ induces the Δ_1-topology on C while making it complete.

Note that (25) is the "bitopological analogue" of the well known fact that G_δ-sets are homeomorphic to complete metric spaces.

An immediate corollary of (25) is the following:

(26) If C is a strong w^*-G_δ in $\bar{C}^*$ and if the latter is w^*-metrizable, then there exists a w^*-lower semi-continuous and norm Lipschitz metric Δ on $\bar{C}^*$ that induces the w^*-topology on C while making it complete.

Before proceeding with the proofs, we note that if Y is a separable Banach space, then all w^*-compact subsets of Y^* are w^*-metrizable. Moreover, if we let T be a compact dense-range operator from ℓ_2 into Y, then $T^*:Y^* \longrightarrow \ell_2$ is a weak* to norm homeomorphism on the w^*-compact subsets of Y^*. In this case w^*-G_δ (resp w^*-H_δ) bounded subsets of Y^* can be identified with G_δ (resp H_δ) subsets of ℓ_2 where it can be more convenient to work. The ℓ_2-norm then defines a metric δ on Y^* that induces the w^*-topology on the bounded subsets and whose balls are w^*-compact and convex.

In most results mentioned above, the only w^*-metrizability assumption made, is on the w^*-closure of the set in question. In many of the proofs we shall find it convenient to assume that the predual is separable. Since this is not the case in general, the following lemma will often allow us to reduce the assumption of "local w^*-metrizability" to the case where the predual is separable.

Lemma 0: Let Y be a Banach space and let C be a subset of Y^* such that $\bar{C}^*$ is w^*-compact and w^*-metrizable. There exists then a separable subspace Z of Y such that if $i:Z \longrightarrow Y$ denotes the canonical embedding we have:

a) The adjoint map $i^*:Y^* \longrightarrow Z^*$ acting as an isometry and a w^*-homeomorphism on $\bar{C}^*$.

b) For any convex subset D of C and for any non-empty w^*-open slice S of D, there exists a non-empty w^*-open slice S_1 of $i^*(D)$ such that $S_1 \subseteq i^*(S)$.

Proof: To insure that i^* is a w^*-homeomorphism on $\bar{C}^*$, it is enough to construct a separable subspace Z_0 of Y that separates the points of $\bar{C}^*$. For that, let d be a metric on $\bar{C}^*$ that determines the w^*-topology and consider for each n the set $F_n = \{(x^*,y^*) \in \bar{C}^* \times \bar{C}^*;\ d(x^*,y^*) \geq \frac{1}{n}\}$. Use now the compactness of $\bar{C}^*$ to cover the diagonal of $\bar{C}^* \times \bar{C}^*$ with open sets of the form $\Omega_n = \bigcup_{m=1}^{M(n)} U_{n,m}$ where $\Omega_n \cap F_n = \emptyset$ for each n and where each $U_{n,m}$ is an elementary w*-open set in Y*. It is clear that the countable number of functionals in Y that determine the w*-open sets $(U_{n,m})$ separate the points of $\bar{C}^*$. The space Z_0 can be taken to be the separable subspace of Y generated by those functionals.

To insure that i* is an isometry on $\bar{C}^*$, we construct by transfinite induction an increasing family of subspaces of Y in the following way: start with Z_0 and if $\alpha = \beta + 1$, let ϕ_β be the function on $\bar{C}^* \times \bar{C}^*$ defined by $\phi_\beta\,(x^*,y^*) = \sup\{|(x^* - y^*,z)|;z \in \mathrm{Ball}(Z_\beta)\}$.

If now $\phi_\beta(x^*,y^*) < \|x^* - y^*\|$ for some (x^*,y^*) in $\bar{C}^* \times \bar{C}^*$, we can find $z_\beta \in Y$ such that if $Z_{\beta+1} = \overline{\mathrm{linspan}(Z_\beta \cup \{z_\beta\})}$ then $\phi_\beta(x^*,y^*) < \phi_{\beta+1}(x^*,y^*)$.

If α is a limit ordinal set $Z_\alpha = \overline{\bigcup_{\beta<\alpha} Z_\beta}$.

Note now that (ϕ_α) is an increasing family of w*-lower semi-continuous functions on the polish space $\bar{C}^* \times \bar{C}^*$ hence there exists $\gamma < \Omega$ (the first uncountable ordinal) such that $\phi_{\gamma+1} = \phi_\gamma$. It follows that Z_γ is separable and that $\phi_\gamma(x^*,y^*) = \|x^* - y^*\|$ for all (x^*,y^*) in $\bar{C}^* \times \bar{C}^*$.

To prove b) assume $x \in S$, hence $x \notin \overline{D \setminus S}^*$ and $i^*(x) \notin \overline{i^*(D) \setminus i^*(S)}^*$, since i* is a w*-homeomorphism on $\bar{C}^*$. It is now enough to use the Hahn-Banach theorem to separate $i^*(x)$ from $\overline{i^*(D) \setminus i^*(S)}^*$ and obtain the w*-open slice S_1 contained in $i^*(S)$ and containing $i^*(x)$. Q.E.D.

We tried to make this paper as self-contained as possible. However, we might have assumed inadvertently that some notions are standard and we refer the interested reader to the books of Diestel-Uhl [10] and Bourgin [4] for more details about the unexplained ones.

We are grateful to I. Ekeland for very fruitful conversations during the preparation of this paper. We would like also to thank Wendy Spaxman from the staff of the department of mathematics at U.B.C. for her excellent job in typing this manuscript.

January 1986.

I. THE EXTREMAL STRUCTURE OF CONVEX w^*-H_δ SETS:

Let Y be a Banach space and let C be a convex bounded subset of Y*. We shall say that C has the *-R.N.P (resp the R.N.P) if for every probability space $(\Omega,\mathcal{F},P)$ and bounded linear operator $T:L^1(P) \longrightarrow Y^*$ with $Tf \in C$ for all f in the positive sphere of $L^1(P)$, there is a w^*-measurable (resp a strongly measurable) function $\phi:\Omega \longrightarrow C$ with $Tf = w^*\text{-}\int \phi f dP$ (resp $Tf = \int \phi f dP$) for all f in $L^1(P)$.

It is easy to see that if the set C is norm separable then the two notions defined above are equivalent.

We shall say that C is w^*-martingale-compact (resp martingale-compact) if given a C-valued martingale (f_n) with f_n finite-valued for each n, then (f_n) w^*-converges (resp norm-converges) almost everywhere to a C-valued function f.

Note that if Y is separable, it is well known that C-valued vector measures (resp martingales) have w^*-densities (resp w^*-limits) that are almost everywhere valued in $\bar{C}^*$. If Y is not separable but $\bar{C}^*$ is w^*-metrizable then the same conclusions hold by a straightforward application of Lemma 0.

Given a subalgebra $\mathcal{A}$ of $\mathcal{F}$, we let $\sigma(\mathcal{A})$ denote the smallest σ-algebra of sets containing $\mathcal{A}$. We recall now a concept introduced by H. P. Rosenthal [25]:

The set C is said to be L^1-convex provided for all probability spaces $(\Omega,\mathcal{F},P)$, subalgebras $\mathcal{A}$ of $\mathcal{F}$ with $\sigma(\mathcal{A}) = \mathcal{F}$ and bounded linear operators $T:L^1(P) \longrightarrow Y^*$ with $Tf \in C$ for all $\mathcal{A}$-measurable functions f in the positive unit sphere $\mathcal{P}(P)$ of $L^1(P)$ then $Tf \in C$ for all $f \in \mathcal{P}(P)$.

The set C is said to be σ-convex if for all sequences (c_i) in C and all sequences of positive reals (λ_i) such that $\sum_{i=1}^{\infty} \lambda_i = 1$ we have $\sum_{i=1}^{\infty} \lambda_i c_i \in C$.

We recall the following result proved by H. P. Rosenthal in [25]: A bounded convex set C is L^1-convex if and only if it is σ-convex and line-closed.

Suppose now C is a w*-G_δ set in $\bar{C}^*$ and that the latter is w*-metrizable. It is well known [19] that there exists then a metric Δ on $\bar{C}^*$, that induces the w*-topology on C in such a way that the metric space (C,Δ) is complete, and that for each x in $\bar{C}^*$, the function $\Delta(x,\cdot)$ is w*-lower semi-continuous on $\bar{C}^*$. Throughout this section, we shall denote by (C,Δ) any w*-G_δ set in $\bar{C}^*$ equipped with a metric Δ on $\bar{C}^*$ satisfying the above mentioned properties. It will be clear from what follows that most of the results that we establish for w*-G_δ sets do not depend on the choice of such a metric.

Suppose now that F is a subset of a w*-G_δ set (C,Δ), it is clear that F is w*-relatively closed in C if and only if it is Δ-closed. In this case, F is also a w*-G_δ set in $\bar{C}^*$. The following lemma shows that the same stability property holds for w*-H_δ sets.

<u>Lemma I.1</u>: Let C be a bounded subset of a dual Banach space Y*.

a) If $\bar{C}^*$ is w*-metrizable then for every relatively w*-closed subset F of C we have: $\bar{C}^* \setminus F = (\bar{C}^* \setminus C) \cup \bigcup_n (K_n \cap \bar{C}^*)$ where each K_n is w*-compact convex with $d(K_n,F) > 0$.

In particular F is a w*-H_δ (resp a strong w*-H_δ) in $\bar{F}^*$, or equivalently in $\bar{C}^*$ whenever C is a w*-H_δ (resp a strong w*-H_δ) in $\bar{C}^*$.

b) If C is norm separable, then the same conclusion holds for every relatively norm-closed subset F of C.

<u>Proof</u>: We start with b). For each x in Y*, denote by r(x) the distance d(x,F) of x to F and let B(x) be the open ball in Y* centered at x and of

radius equal to $\frac{r(x)}{2}$. Note that $C \setminus F = \bigcup_{x \in C\setminus F} B(x) \cap C$ and each $B(x) \cap C$ is a relatively open set in C which is separable, hence there exists a countable subfamily (x_n) in $C \setminus F$ such that $C \setminus F = \bigcup_n B(x_n) \cap C$. We claim that the sets defined for each n by $K_n = \overline{B(x_n)}$ verify the claim of the lemma. Indeed, each K_n is w*-compact and convex. Moreover, for each n we have that $d(K_n, F) \geq \frac{r(x_n)}{2} > 0$ hence $F \cap (\bigcup_n K_n) = \emptyset$ and $\bar{C}^* \setminus F = (\bar{C}^* \setminus C) \cup (\bigcup_n (K_n \cap \bar{C}^*))$.

To prove a) note first that by lemma 0 we can assume that Y is separable. Let now T be a compact dense range operator from ℓ_2 into Y. Then T* is one-to-one, norm compact and acts as an homeomorphism between $(\bar{C}^*, w^*)$ and $(T^*(\bar{C}^*), \| \ \|)$. Note that T*(F) is then a relatively norm closed subset of T*(C) which is norm-separable in ℓ_2. Part b) of the Lemma then applies and we get that

$$T^*(\bar{C}^*) \setminus T^*(F) = (T^*(\bar{C}^*) \setminus T^*(C)) \cup (\bigcup_n (K_n \cap T^*(\bar{C}^*))$$

where each K_n is convex compact in ℓ_2 such that $d(K_n, T^*(F)) > 0$. Write now $\bar{C}^* \setminus F = (\bar{C}^* \setminus C) \cup \bigcup_n (T^{*-1}(K_n \cap T^*(\bar{C}^*))$ which clearly verify the claim in part a).

If now Δ is any metric on the w*-closure of a bounded subset C of Y*, we shall say that C is <u>w*-Δ-dentable</u> if every relatively Δ-closed subset of C has non-empty w*-open slices of arbitrarily small Δ-diameter. We shall say that C is <u>completely w*-Δ-dentable</u> if for every relatively Δ-closed subset F of C, every non-empty w*-open slice S of F and every $\varepsilon > 0$, there exists a non-empty w*-open slice S_1 of F with $S_1 \subseteq S$ and $\Delta\text{-diam}(S_1) \leq \varepsilon$.
If Δ is the norm of Y* then we shall simply say that C is <u>w*-dentable</u> (resp <u>completely w*-dentable</u>). The following proposition gives the connection between dentability and H_δ-sets.

Proposition I.2: Let C be a convex bounded subset of a dual space Y^* such that $\overline{C}^*$ is w*-metrizable. Suppose that C is w*-Δ-dentable for a metric Δ on $\overline{C}^*$ verifying the following properties:

(i) For each x in $\overline{C}^*$, the function $\Delta(x,\cdot)$ is w*-lower semi-continuous.

(ii) The metric Δ induces on $\overline{C}^*$ a topology which is finer than the w*-topology.

(iii) (C,Δ) is complete and separable

Then C is a w*-H_δ set in $\overline{C}^*$.

Proof: Fix $\varepsilon > 0$ and define inductively a decreasing family of relatively Δ-closed convex subsets of C in the following manner:

(i) $F_0 = C$

(ii) If $\alpha = \beta + 1$ and $F_\beta \neq \phi$ use the w*-Δ-dentability to find a w*-open half-space H_β such that $H_\beta \cap F_\beta \neq \phi$ and $\Delta\text{-diam}(H_\beta \cap F_\beta) \leq \varepsilon$. Then set $F_\alpha = F_\beta \setminus H_\beta$.

(iii) If α is a limit ordinal, let $F_\alpha = \bigcap_{\beta<\alpha} F_\beta$.

Since (C,Δ) is a Polish space, there is $\gamma_\varepsilon < \Omega$ (the first uncountable ordinal) such that $F_{\gamma_\varepsilon} = \phi$. Let K_α be the w*-closure of F_α in Y^*. Note that

$$C \subseteq \bigcap_{\alpha \leq \gamma_\varepsilon} \left(K_\alpha \cup \bigcup_{\beta<\alpha} H_\beta\right)$$

Moreover, if x belongs to the right-hand side, then there exists $\beta < \gamma_\varepsilon$ such that $x \in K_\beta \cap H_\beta$. Let now (x_j) be a sequence in F_β such that $x = \text{w*-}\lim_j x_j$. Since H_β is w*-open, we have that for a large enough j, $x_i \in F_\beta \cap H_\beta$ for all $i \geq j$. Since $\Delta(\cdot,x_j)$ is w*-lower semi-continuous on $\overline{C}^*$, we get

$$\Delta(x,x_j) \leq \varliminf_i \Delta(x_i,x_j) \leq \varepsilon \quad \text{hence} \quad \Delta(x,C) \leq \varepsilon.$$

It follows that if we repeat the construction for each $\varepsilon = \frac{1}{n}$ we would get (since C is Δ-closed in $\bar{C}^*$) that:

$$C = \bigcap_n \bigcap_{\alpha<\gamma_n} (K_{\alpha,n} \cup \bigcup_{\beta<\alpha} H_{\beta,n})$$

Use now Lemma 0 to find a separable subspace Z of Y and an operator $i^*:Y^* \to Z^*$ which acts as a w*-homeomorphism on C*. Note that each $K_{\alpha,n}$ is a w*-compact subset of $\bar{C}^*$, hence $i^*(K_{\alpha,n})$ is a w*-H_δ in Z*. It follows that $\bar{C}^* \setminus C = \bigcup_n H_n$ where each H_n is w*-compact and convex in Y* and C is then a w*-H_δ set in $\bar{C}^*$.

<u>Theorem I.3</u>: Let C be a convex bounded subset of a dual space Y* such that $\bar{C}^*$ is w*-metrizable. Suppose (C,Δ) is a w*-G_δ in $\bar{C}^*$. The following properties are then equivalent:

1) C is a line-closed w*-H_δ subset of $\bar{C}^*$.

2) C is a line-closed *-R.N.P set.

3) C is w*-martingale-compact.

4) C is completely w*-Δ-dentable.

<u>Proof</u>: 1) => 2) This is an adaptation of a proof of Edgar-Wheeler [11]. Let (Ω,Σ,μ) be a probability space and let $F:\Sigma \longrightarrow Y^*$ be a vector measure such that $\frac{F(A)}{\mu(A)} \in C$ for each $A \in \Sigma$ with $\mu(A) > 0$. There exists then a w*-measurable function $\phi:\Omega \longrightarrow \bar{C}^*$ such that for each $E \in \Sigma$ we have

$$F(E) = w^*\text{-}\int_E \phi(t)\,d\mu(t)$$

To show that ϕ has almost all its values in C, write $\bar{C}^* \setminus C = \bigcup_n K_n$ where each K_n is w*-compact and convex. For each n, the set $D_n = \phi^{-1}(K_n)$ belongs to Σ and if $\mu(D_n) > 0$ we get that $\frac{F(D_n)}{\mu(D_n)} \in C$. On the other hand we have

$\frac{F(D_n)}{\mu(D_n)} = \frac{1}{\mu(D_n)} \cdot \text{w*-}\int_{D_n} \phi(t)d\mu(t)$ which belongs to K_n since the latter is w*-compact and convex. It follows that $\mu(D_n) = 0$ for each n and $\phi(t) \in C$ a.s.

2) => 3) Note first that by Lemma 0, C is linearly homeomorphic to a bounded convex w*-G_δ set in a dual Banach space. By a result of Fremlin-Talagrand [13] such sets are always σ-convex. The hypothesis in 2) then implies, via the result of Rosenthal mentioned above, that C is L^1-convex. Let now (ϕ_n) be a C-valued martingale defined on a probability space (Ω,Σ,μ) with ϕ_n finite-valued for each n, and let $\mathcal{A}_n$ be the algebra of sets depending on $\{\phi_1,\dots,\phi_n\}$ for each n. Let $\mathcal{A} = \bigcup_n \mathcal{A}_n$ and assume without loss of generality that $\Sigma = \sigma(\mathcal{A})$. Let now T be the operator from $L^1(\Omega,\Sigma,\mu)$ into Y* defined by

$$Tf = \lim_n \int f\phi_n d\mu$$

Since C is L^1-convex, the same argument as Rosenthal's [25] shows that $Tf \in C$ for all f in $\mathcal{P}(\mu)$. Since C is a *-R.N.P set, there exists a w*-measurable function ϕ which is almost surely valued in C such that

$$Tf = \text{w*-}\int f\phi d\mu \quad \text{for each } f \text{ in } \mathcal{P}(\mu)$$

On the other hand, if $\tilde{\phi}$ is the w*-limit of the martingale (ϕ_n) we must have $\tilde{\phi} = \phi$ a.s. It follows that C is w*-martingale-compact.

To prove that 3) => 4) we need the following

Lemma I.4: Let (C,Δ) be a bounded convex G_δ subset of ℓ_2. If C is martingale compact, then for each non-empty open slice S of C and every $\varepsilon > 0$, there exists a non-empty open slice S_1 of C with $S_1 \subseteq S$ and such that the Δ-diameter of $\overline{S}_1$ is less than ε.

Proof: Let $x^* \in \ell_2$, $\|x^*\| = 1$ and let $\alpha \in \mathbf{R}$ such that $S = \{x^*>\alpha\} \cap C \neq \phi$. Let $V = \{x^* \leq \alpha\} \cap C$. By the Hahn-Banach theorem it is enough to show that there exists $x \in S$ such that $x \notin \overline{\mathrm{conv}}(\{C \setminus B_\Delta(x,\varepsilon)\} \cup V)$ where the closure is taken in ℓ_2. Suppose that such a point does not exist. We shall construct inductively a finitely valued quasi-martingale (Q_n, F_n) on ([0,1), Borel sets, P = Lebesgue measure) in the following way:

(a) Start with $Q_0 = x_0$ where x_0 is any point in S.

(b) If at the nth level $Q_n(\omega) \in V$, we set $Q_{n+1}(\omega) = Q_n(\omega)$.

(c) If $q = Q_n(\omega) \notin V$, we consider the finite set $L_n = \{E(Q_n|F_j)(\omega);$ $0 \leq j \leq n; \omega \in \Omega\}$ in C and we choose α_n such that: $\forall x \in L_n$, $\forall z \in C$, $\|x-z\| \leq \alpha_n \Rightarrow \Delta(x,z) \leq 2^{-n}$.

Let now $y = \sum_{i=1}^{N} \theta_i y_i$, $\theta_i \geq 0$, $\sum_{i=1}^{N} \theta_i = 1$ such that $y_i \in \{C \setminus B_\Delta(q,\varepsilon)\} \cup V$ and $\|q-y\| \leq \alpha_n$. We split the atom $\{Q_n = q\}$ of F_n into N disjoint subsets $(A_i)_{i=1}^N$ such that $P(A_i) = \theta_i P(Q_n=q)$ and we give Q_{n+1} the value y_i on each A_i.

Note that for each n and each ω, $\|E(Q_{n+1}|F_n)(\omega) - Q_n(\omega)\| \leq \alpha_n$, hence $\forall n,k$, $\|E(Q_{n+k+1}|F_n)(\omega) - E(Q_{n+k}|F_n)(\omega)\| \leq \alpha_{n+k}$. Since $E(Q_{n+k}|F_n)$ takes its values in L_{n+k}, we get that

$$\Delta(E(Q_{n+k+1}|F_n)(\omega), E(Q_{n+k}|F_n)(\omega)) \leq 2^{-n-k}.$$

It follows that for each n, $E(Q_{n+k}|F_n)$ converges when $k \to \infty$ to a random variable M_n valued in C since (C,Δ) is complete. Moreover, the sequence (M_n)

is a martingale which converges to M_∞. By the assumption on C, $M_\infty \in C$ a.s.

Suppose now that the α_n's are chosen in such a way that $\delta_0 = \sum_{j=0}^{\infty} \alpha_j < x^*(x_0) - \alpha$. We get that for each n and each ω, $\|Q_n(\omega) - M_n(\omega)\| \leq \sum_{i=n}^{\infty} \alpha_i$ from which follows that (Q_n) converges to the same limit M_∞.

On the other hand, we have $\|\int Q_n dP - x_0\| \leq \sum_{i=0}^{n} \alpha_i$ for each n, hence $\|\int M_\infty dP - x_0\| \leq \delta_0$ and $\int x^*(M_\infty) dP \geq x^*(x_0) - \delta_0 > \alpha$. It follows that $P(M_\infty \in S) > 0$. But for an ω such that $M_\infty(\omega) \in S$ we have necessarily that $Q_n(\omega) \in S$ for each n. This means that $\Delta(Q_n(\omega), Q_{n+1}(\omega)) \geq \varepsilon$ for each n which clearly contradicts the convergence of (Q_n) in C.

To prove 3) => 4) of Theorem I.3 suppose that (C,Δ) is a bounded convex w*-G_δ subset of $\bar{C}^*$ which is also w*-martingale compact. By Lemma 0, we can assume without loss of generality that Y is separable. Let T be a dense range operator from ℓ_2 into Y. Note that $T^*(C)$ is a convex bounded subset of ℓ_2 that verifies the hypothesis of Lemma I.4 with the metric Δ' on $\overline{T^*(C)}$ defined by $\Delta'(T^*c_1, T^*c_2) = \Delta(c_1, c_2)$ for each c_1, c_2 in $\bar{C}^*$. Apply now Lemma I.4 to $T^*(C)$ to obtain the claimed result for C.

To prove 4) => 1) in Theorem I.3, note that if C is completely w*-Δ-dentable then by Proposition I.2 it is a w*-H_δ set in $\bar{C}^*$. Moreover, every relatively Δ-closed subset of C is contained in the w*-closed convex hull of its extreme points (see the proof of Corollaries I.5 and I.6). It follows that if the open segment $[x,y[$ is in C and $x \neq y$ then x cannot be the only extreme point of $[x,y[$ hence $y \in C$ and C is line-closed.

<u>Corollary I.5</u>: Let C be a bounded convex line-closed subset of ℓ_2. If C is an H_δ-set then it is contained in the closed convex hull of its denting points.

Proof: Since C is G_δ, we let Δ be a metric that induces the norm- topology on C while making it complete. Let S be a non-empty slice of C. Apply Lemma I.4 to obtain a decreasing sequence (S_n) of slices of C that are contained in S and such that the Δ-diameter of $\overline{S}_n$ goes to zero. It follows from the Δ-completeness of C that there exists x in C such that $\{x\} = \bigcap_n \overline{S}_n$. Clearly x is a denting point of C which is contained in S, the rest is an immediate consequence of the above and the Hahn-Banach theorem.

Corollary I.6: Let C be a bounded convex line-closed subset of a dual space Y* such that $D = \overline{C}^*$ is w*-metrizable. If C is a w*-H_δ set in D then D is the w*-closed convex hull of its extreme points that are in C. Moreover, if E is a relatively w*-closed subset of C such that $D = \overline{\mathrm{conv}}^*(E)$, then E must contain the w*-strongly extreme points of C.

Proof: Again, by Lemma 0, we can assume without loss of generality that Y is separable. Let now T be a dense range operator from ℓ_2 into Y. Apply Corollary I.5 to the set T*(C) and note that the denting points of T*(C) correspond to the w*-strongly extreme points of C which are exactly the extreme points of D that are in C.

Suppose now x is a w*-strongly extreme point of C which is not in E. There exists then a w*-open neighborhood V such that $x \in V \cap C$ and $V \cap E = \emptyset$. Find now a w*-open slice S of C such that $x \in S \subseteq V \cap C$, thus $x \notin \overline{\mathrm{conv}}^*(E)$; a contradiction.

For the definitions of G_δ-embeddings and H_δ-embeddings we refer to [14].

Corollary I.7: Let X be a Banach space such that there exists a G_δ-embedding T from X into ℓ_2. Then X has the Radon-Nikodym property if and only if T is an H_δ-embedding.

Proof: If T was an H_δ-embedding then X is separable and has the Radon-Nikodym property by Theorem III.3 of [14]. The same theorem asserts that there exists an H_δ-embedding from X into ℓ_2 whenever X has the Radon-Nikodym property. The interest in this corollary lies in the fact that the same operator T is actually an H_δ-embedding. To show that, note that $T(B_X)$ is an L^1-convex bounded G_δ in ℓ_2. Moreover, it is martingale-compact since B_X has the R.N.P. and T is one-to-one. It follows from Theorem I.3 that $T(B_X)$ is an H_δ-subset of ℓ_2.

We now deal with w^*-H_δ sets having the Radon-Nikodym property. Theorem I.8 below extends the results of C. Stegall [26] on w^*-compact convex sets. We shall also see in section IV that it also contains the results of J. Bourgain [1] on subsets of duals of spaces not containing a copy of ℓ_1.

Theorem I.8: Let C be a bounded convex and line-closed subset of a dual space Y^* such that $\bar{C}^*$ is w^*-metrizable. If C is a w^*-H_δ in $\bar{C}^*$ then the following conditions are equivalent:

1) C is w^*-dentable.

2) C is separable.

3) C has the Radon-Nikodym property.

4) C contains no δ-tree for any $\delta > 0$.

5) C is completely w^*-dentable.

Proof: 1) => 2) If C is not separable, then there exists $\varepsilon > 0$ and an uncountable subset D of C such that the distance between any two points of D exceeds ε. Since $\bar{C}^*$ is w^*-compact, D contains an uncountable subset E that is w^*-dense in itself. Let $F = \bar{E}^*$. It follows that for every w^*-open set V with $V \cap F \neq \phi$ we have $\operatorname{diam}(V \cap F) > \varepsilon$. Hence C is non w^*-dentable.

Note that if C is norm closed and w^*-dentable then Proposition I.2, applied to the metric induced by the norm, gives that C is a w*-H_δ in $\bar{C}^*$.

2) => 3) Since C is a bounded convex line-closed w*-H_δ set in $\bar{C}^*$, Theorem I.3 gives that C has the *-R.N.P. The separability gives then that C has the R.N.P.

3) => 4) is clear. The rest of this section is devoted to the proof of 4) => 5). The following is a version of a Lemma of Bourgain [1].

Lemma I.9: Let C be a bounded convex subset in ℓ_2 and let B be a closed bounded convex symmetric subset of ℓ_2 such that $C \subseteq B$. Let E be a subset of C such that $C \subseteq \overline{\mathrm{conv}}(E)$. If S is a non-empty open slice of C such that $E \cap S \subseteq u + \frac{\tau}{2} B$ for some u in ℓ_2 and some $\tau > 0$, then there exists a non-empty open slice $\tilde{S}$ of C such that $\tilde{S} \subseteq S$ and $\tilde{S} \subseteq u + \tau B$.

Proof: Suppose $S = C \cap \{x^* > \alpha\}$ for some x^* in ℓ_2 and $\alpha \in \mathbf{R}$. Let $\beta = \sup x^*(C)$ and choose $0 < \sigma < \frac{\tau}{4}(\beta-\alpha)$ in such a way that $\tilde{S} = C \cap \{x^* > \beta - \sigma\}$ is non-empty and $\tilde{S} \subseteq S$. Set $E_1 = E \cap S$ and $S_2 = C \cap \{x^* \leq \alpha\}$. We have:

$$C \subseteq \overline{\mathrm{conv}}(E) \subseteq \overline{\mathrm{conv}}(E_1 \cup S_2) \subseteq \overline{\mathrm{conv}}\ (\overline{\mathrm{conv}}(E_1) \cup \bar{S}_2) = \mathrm{conv}(\overline{\mathrm{conv}}(E_1) \cup \bar{S}_2).$$

Let $x \in \tilde{S}$. Write $x = (1-\lambda)y + \lambda z$ with $0 < \lambda \leq 1$, $y \in \overline{\mathrm{conv}(E_1)}$ and $z \in \bar{S}_2$. We get that $x^*(y) \leq \beta$ and $x^*(z) \leq \alpha$ hence:

$$\beta - \sigma < x^*(x) \leq (1-\lambda)\beta + \lambda\alpha \quad \text{and} \quad \lambda \leq \frac{\sigma}{\beta-\alpha} < \frac{\tau}{4}. \quad \text{Thus}$$

$x - y = \lambda(z-y) \in \frac{\tau}{4}(B-B) = \frac{\tau}{2} B$. It follows that

$$\tilde{S} \subseteq \overline{\mathrm{conv}}(E_1) + \frac{\tau}{2} B \subseteq (u + \frac{\tau}{2} B) + \frac{\tau}{2} B = u + \tau B.$$

Let now C be a bounded convex and line-closed H_δ subset of ℓ_2 and let B as in the hypothesis of Lemma I.9. We shall denote by E the set of denting points of C. Note that by Corollary I.5, $C \subseteq \overline{\text{conv}}(E)$.

Lemma I.10: Let $\tau > 0$ and let S be a non-empty open slice of C such that no slice $\tilde{S}$ of C with $\tilde{S} \subseteq S$ is contained in a set of the form $u + \tau B$. There exists then a dyadic martingale (M_n) valued in S such that for each n, $M_{n+1} - M_n \notin \frac{\tau}{4} B$.

Proof: Let Δ be a metric on C that induces on it the ℓ_2-topology while making it complete. Let $T = \bigcup_{n=0}^{\infty} \{-1,+1\}^n$ be the usual dyadic tree.

To build the dyadic martingale, we start with the following inductive construction:

Suppose $S = C \cap \{x^* > \alpha\}$ where $x^* \in \ell_2$ and $\alpha \in \mathbf{R}$. Choose $\beta > \alpha$ such that $S_\phi = C \cap \{x^* > \beta\}$ is non empty. By Lemma I.9, there exists two points $x_\varepsilon \in E \cap S_\phi$ ($\varepsilon = \pm 1$) such that $x_1 - x_{-1} \notin \frac{\tau}{2} B$. By the Hahn-Banach theorem, find $x^*_\phi \in \ell_2$ such that $x^*_\phi(x_1 - x_{-1}) > \frac{\tau}{2}$ and $\sup x^*_\phi(B) < 1$. Set $\rho_\phi = x^*_\phi\left(\frac{x_1+x_{-1}}{2}\right)$ and $U_\varepsilon = \{\varepsilon x^*_\phi > \varepsilon \rho_\phi + \frac{\tau}{4}\}$ for $\varepsilon = \pm 1$. Note that $x_\varepsilon \in U_\varepsilon \cap S_\phi$ which is open. Since $x_\varepsilon \in E$, there exists two slices S_ε of C, $S_\varepsilon \subseteq S_\phi$ such that $x_\varepsilon \in S_\varepsilon \subseteq U_\varepsilon \cap S_\phi$ and $\text{diam}_2(S_\varepsilon) \leq \alpha_1$ where α_1 is such that

$$\forall z \in C, \ \left\|z - \frac{x_1+x_{-1}}{2}\right\| \leq \alpha_1 \Rightarrow \Delta\left(z, \frac{x_1+x_{-1}}{2}\right) \leq \frac{1}{2}.$$

For the general step, suppose we had x^*_t in ℓ_2, ρ_t in $\mathbf{R}$, $x_{t,\varepsilon}$ in C and $S_{t,\varepsilon}$ for $0 \leq |t| \leq n-1$, $\varepsilon = \pm 1$ and $(\alpha_j)_{1 \leq j \leq n}$, $\alpha_j > 0$ such that

a) $x_{t,\varepsilon} \in E \cap S_{t,\varepsilon}$ where $S_{t,\varepsilon}$ is an open slice of C.

b) If $L_{j+1} = \text{conv}\{x_{t,\varepsilon}; |t| = j, \varepsilon = \pm 1\}$ for $0 \leq j \leq n-1$, then

$$\forall z \in C, \forall x \in L_{j+1}, \|z - x\| \leq \alpha_{j+1} \Rightarrow \Delta(z,x) \leq 2^{-j-1}.$$

c) $\sup x_t^*(B) < 1$ and $x_t^*(x_{t,-1}) < \rho_t' - \frac{\tau}{4} < \rho_t' + \frac{\tau}{4} < x_t^*(x_{t,1})$

d) $S_{t,\varepsilon} \subseteq S_t \cap U_{t,\varepsilon}$ where $U_{t,\varepsilon} = \{\varepsilon x_t^* > \varepsilon\rho_t' + \frac{\tau}{4}\}$ and $\text{diam}_2(S_{t,\varepsilon}) \leq \alpha_{|t|+1}$.

Let $t \in T$ such that $|t| = n$. Since S_t is an open slice of C, use Lemma I.9 to choose two points $x_{t,\varepsilon}(\varepsilon=\pm 1)$ such that $x_{t,\varepsilon} \in E \cap S_t$ and $x_{t,1} - x_{t,-1} \notin \frac{\tau}{2} B$.

Let $L_{n+1} = \text{conv}\{x_{t,\varepsilon}; |t| = n;\ \varepsilon = \pm 1\}$ and choose $\alpha_{n+1} > 0$ such that: $\forall z \in C,\ \forall x \in L_{n+1},\ \|x-z\| \leq \alpha_{n+1} \Rightarrow \Delta(z,x) \leq 2^{-n-1}$.

Since $x_{t,1} - x_{t,-1} \notin \frac{\tau}{2} B$ we can find x_t^* such that $\sup x_t^*(B) < 1$ and $x_t^*(x_{t,1} - x_{t,-1}) > \frac{\tau}{2}$. Set $\rho_t = x_t^*(\frac{x_{t,1}+x_{t,-1}}{2})$, then $x_{t,\varepsilon} \in S_t \cap U_{t,\varepsilon}$ which is open and since $x_{t,\varepsilon} \in E$, we can find two slices $S_{t,\varepsilon}$ of C such that

$$x_{t,\varepsilon} \in S_{t,\varepsilon} \subseteq S_t \cap U_{t,\varepsilon} \text{ and } \text{diam}_2(S_{t,\varepsilon}) \leq \alpha_{n+1}$$

This finishes the inductive step.

Consider now the probability space $\Omega = \{-1;+1\}^{\mathbf{N}}$ equipped with its canonical probability measure and its dyadic sub-σ-fields $(\mathcal{F}_n)$. To each t in T, we associate the atom A_t in $\mathcal{F}_{|t|}$ defined by $A_t = \{w = (\varepsilon_n);\ (\varepsilon_1,\ldots\varepsilon_{|t|}) = t\}$. We define a sequence of adapted random variables $(Q_n, \mathcal{F}_n)$ valued in C in such a way that $Q_0 = x_0$ where x_0 is any point in S_ϕ and if $|t| = n$ we let $Q_{|t|} = x_t$ on the atom A_t.

If $|t| = n$, note that $x_{t,\varepsilon} \in S_{t,\varepsilon} \cap S_t$ and $\text{diam}_2(S_t) \leq \alpha_n$, hence for each $n \geq 1$, $\forall w$, $\|Q_{n+1}(w) - Q_n(w)\| \leq \alpha_n$. It follows that for each $k \leq n$, $\forall w$, $\|E(Q_{n+1}|\mathcal{F}_k)(w) - E(Q_n|\mathcal{F}_k)(w)\| \leq \alpha_n$. Since $E(Q_{n+1}|\mathcal{F}_k)(\omega) \in C$ and $E(Q_n|\mathcal{F}_k)(\omega) \in L_n$ we get that

$$\Delta(E(Q_{n+1}|\mathcal{F}_k)(w),\ E(Q_n|\mathcal{F}_k)(w)) \leq 2^{-n}.$$

It follows that $E(Q_n|\mathcal{F}_k)$ converges in C when $n \to \infty$ and that $M_k = \lim_n E(Q_n|\mathcal{F}_k)$ is a dyadic martingale represented by a dyadic tree $\{z_t; t \in T\}$ in C where z_t is the constant value of $M_{|t|}$ on the atom A_t.

Let now t in T, $k = |t| + 1$, w in $A_{t,\varepsilon}$, $n \geq k$. We have $Q_n(w) \in S_{t,\varepsilon} \cap \{\varepsilon x^*_t > \varepsilon \rho_t + \frac{\tau}{4}\}$ hence $E(Q_n | F_k)(w) \in S_{t,\varepsilon}$ and by taking the limit we get that $\varepsilon x^*_t(z_{t,\varepsilon}) \geq \varepsilon \rho_t + \frac{\tau}{4}$.

It follows that $x^*_t(z_{t,1} - z_{t,-1}) \geq \frac{\tau}{2}$ and that $z_{t,1} - z_{t,-1} \notin \frac{\tau}{2} B$. This finishes the proof of Lemma I.10.

To prove 4) => 5) of Theorem I.8, suppose first that F is a convex line-closed subset of C that is a w^*-H_δ in $\bar{C}^*$ and suppose there exists a w^*-open slice S of F and $\tau > 0$ such that every w^*-open slice of F contained in S has a diameter greater than τ. Again, by Lemma 0, we can assume Y separable and we let $T: \ell_2 \to Y$ be a dense range operator such that $T^*: Y^* \to \ell_2$ is one-to-one. Note that $T^*(F)$ verifies the hypothesis of lemma I.10 with $B = T^*(B_{Y^*})$. The conclusion of the same lemma gives that F contains a $\frac{\tau}{4}$ - tree which is a contradiction.

To prove that C is strongly w^*--dentable, it remains to show that the same holds for every relatively-norm closed convex subset F of C. For that, it is enough in view of Lemma I.1 to show that C is norm separable.

If C is not separable, then there exists $\tau > 0$ such that C cannot be written as a countable union of subsets with diameter less than τ. Let now S_1 be a w^*-open slice of C such that $\operatorname{diam}(S_1) \leq \tau$. Note that $C_1 = C \setminus S_1$ is a relatively w^*-closed convex subset of C. Moreover, it is line-closed and a w^*-H_δ set in $\bar{C}^*$. By transfinite induction, we can define a decreasing family of such sets (C_α) and a family of w^*-open slices S_α of C_α such that $C_{\alpha+1} = C_\alpha \setminus S_\alpha$ and $\operatorname{diam}(S_\alpha) \leq \tau$. For a limit ordinal α, we take $C_\alpha = \bigcap_{\beta<\alpha} C_\beta$. Note that (C_α) is a decreasing family of w^*-relatively closed subsets of C which is polish in the w^*-topology. It follows that there exists $\gamma < \Omega$ (the first uncountable ordinal) such that $C_\gamma = \phi$. But this means that $C = \bigcup_{\alpha<\gamma} S_\alpha$

and $\mathrm{diam}(S_\alpha) \leq \tau$ which contradicts the non-separability of C.

<u>Corollary I.11</u>: Let C be a convex bounded line-closed subset of a dual space Y^*, such that $D = \bar{C}^*$ is w*-metrizable. If C is a w*-H_δ in D which contains no δ-tree for any $\delta > 0$, then D is the w*-closed convex hull of the w*-denting points of C. Moreover, if E is a relatively norm closed subset of C such that $D = \overline{\mathrm{conv}}^*(E)$ then E must contain the w*-denting points of C.

<u>Proof</u>: Theorem I.8 gives that C is completely w*-dentable. Let now Δ be a metric on C that induces the w*-topology on C while making it complete. Theorem I.3 gives then that C is completely w*-Δ-dentable. Use now these properties to construct a decreasing sequence (S_n) of w*-slices of C, contained in any given w*-slice S_0 in such a way that the Δ-diameters and the $\|\ \|$-diameters of $(\bar{S}_n^*)$ go to zero. It follows from the Δ-completeness of C that there exists x in C such that $\{x\} = \bigcap_n \bar{S}_n$. Clearly x is a w*-denting point of C which is contained in S_0. The rest is an immediate consequence of the Hahn-Banach theorem.

If now x is a w*-denting point of C which is not in E, find a ball $B(x,\rho)$ which is disjoint from E and let S be a w*-open slice of C such that $x \in S \subseteq C \cap B(x,\rho)$. It follows that $x \notin \overline{\mathrm{conv}}^*(E)$: a contradiction.

Before dealing with the integral representations on w*-H_δ sets we shall study the measurable structure of the extreme points (resp w*-strongly extreme points) of such sets. Note first that Jayne and Rogers [17] gave an example of a closed bounded convex subset C of ℓ_1 such that Ext(C) is not a Borel set. Since ℓ_1 is a separable dual, Lemma I.1.b gives that C is necessarily a w*-H_δ set in ℓ_1. We shall see in the following proposition that the set (denoted $\xi(C)$) of w*-strongly extreme points of a w*-H_δ set C has a nicer measurable structure.

Proposition I.12: Let C be a convex bounded line-closed subset of a dual space Y* such that $D = \overline{C}^*$ is w*-metrizable. If C is a w*-H_δ in D then:

a) $\xi(C)$ is a w*-H_δ set in D.

b) Ext(C) is a w*-co-analytic set in D.

Proof: Since $\xi(C) = C \cap \mathrm{Ext}(D)$, it is enough to prove that Ext(D) is a w*-H_δ set. For that, let $(y_n)_n$ be a countable sequence in B_Y that separates the points of D and consider the w*-continuous strictly convex function ϕ on D defined by $\phi(x^*) = \sum_{n=0}^{\infty} 2^{-n}|y_n(x^*)|^2$. Let $\hat{\phi}(x^*) = \inf\{\ell(x^*),\ \ell$ w*-continous and affine on D with $\ell \geq \phi$ on D$\}$. It is well known [6] that $\hat{\phi}$ is concave, w*-upper semi-continuous and that $\mathrm{Ext}(D) = \{x^* \in D;\ \hat{\phi}(x^*) = \phi(x^*)\}$. Note now that $D \setminus \mathrm{Ext}(D) = \bigcup_n K_n$ where each $K_n = \{x^* \in D; (\hat{\phi}-\phi)(x) \geq \frac{1}{n}\}$ is w*-compact and convex.

b) Let Δ be the diagonal in $C \times C$. It is relatively w*-closed in $C \times C$. Note that $C \setminus \mathrm{Ext}(C) = \phi(C \times C \setminus \Delta)$ where ϕ is the w*-continuous function from $C \times C$ to C defined by $(x,y) \longrightarrow \frac{x+y}{2}$. Hence Ext(C) is w*-co-analytic in C (equivalently in D).

The following corollary can be viewed as a convex analogue of the Baire Category theorem.

Corollary I.13: Let D be a convex w*-compact and w*-metrizable subset of a dual space Y*. Let (C_n) be a sequence of convex line-closed w*-dense w*-H_δ subsets of D. then $D = \overline{\mathrm{conv}}^*(\bigcap_n \xi(C_n))$.

Proof: Proposition I.12 and Corollary I.6 give that for each n, $\xi(C_n)$ is a w*-H_δ subset of Ext(D) verifying $D = \overline{\mathrm{conv}}^*(\xi(C_n))$. We get from Krein's theorem that $\overline{\xi(C_n)}^* = \overline{\mathrm{Ext}(D)}^*$. It follows from the Baire theorem that

$\overline{\mathrm{Ext}(D)}^* = \overline{\bigcap_n \xi(C_n)}^*$. Hence $D = \overline{\mathrm{conv}}^*(\mathrm{Ext}(D)) = \overline{\mathrm{conv}}^*(\overline{\bigcap_n \xi(C_n)}^*)$
$= \overline{\mathrm{conv}}^*(\bigcap_n \xi(C_n))$.

To deal with the problem of integral representations on a set C, one needs (as noted by Edgar [4] Corollary 6.3.15) that C be a polish space and that all C-valued martingales converge to a limit in C. If C is now a bounded convex line-closed subset of a dual space Y* such that $\bar{C}^*$ is w*-metrizable and if C is a w*-H_δ in $\bar{C}^*$ then Theorem I.3 gives that finite-valued martingales in C w*-converges almost surely to a C-valued random variable. To insure that all w*-measurable (and not only the finite-valued) martingales in C w*-converge to a C-valued random variable, one needs as noted by H. P. Rosenthal [25] that C be <u>w*-measure convex</u>: that is the (weak*) barycenter of any C-supported w*-Radon measure belongs to C. We do not know if any bounded convex and line-closed w*-H_δ set is necessarily w*-measure convex. However, if we omit the hypothesis of line-closedness, we have the following counterexample due to Weizsaker [30]:

Let $\mathcal{P}$ [0,1] be the set of Radon probability measures on [0,1] and let λ be the Lebesgue measure, then the set $C = \mathcal{P}[0,1] \setminus L_\infty([0,1],\lambda)$ is a bounded convex w*-H_δ in C[0,1]* which is neither line-closed nor w*-measure convex.

An immediate adaptation of Edgar's proof ([4] Theorem 6.2.9) now gives:

<u>Theorem I.14</u>: Let C be a bounded w*-measure convex and line-closed subset of a dual Banach space Y* such that $\bar{C}^*$ is w*-metrizable. If C is a w*-H_δ set in $\bar{C}^*$ then for any x in C, there exists a w*-Radon probability measure μ such that $\mu(\mathrm{Ext}(C)) = 1$ and $x = w^*\text{-}\int t \, d\mu(t)$.

We shall sketch a different proof of this result that exploits the fact that C is linearly homeomorphic to an H_δ-set in ℓ_2.

<u>Proposition I.15</u>: Let C be a bounded measure-convex and line-closed subset of ℓ_2. If C is an H_δ-set then any point x in C is the barycenter of a Radon

probability measure supported by the extreme points of C.

Sketch of proof: Define for each x in C the function

$\rho^2(x) = \sup\{E\|U\|^2;$ U is an ℓ_2-valued random variable such that $E(U) = 0$ and $x + U \in C\}$. Note that $\rho(x) = 0$ if and only if $x \in \mathrm{Ext}(C)$.

Construct now inductively the following martingale: $M_0 = x$ and when $(M_n, \mathcal{F}_n)$ is defined, use the definition of $\rho^2(M_n)$ and standard selection theorems to get M_{n+1} such that $M_n = E_n[M_{n+1}]$ (where E_n denotes the conditional expectation with respect to $\mathcal{F}_n$) and $\rho^2(M_n) - E_n\|M_{n+1} - M_n\|^2 \le 2^{-n}$. Let now M_∞ be the limit of (M_n). Suppose that $\rho^2(M_\infty)$ is not zero with a strictly positive probability and let U be such that $E\|U\|^2 > 0$, $E_\infty(U) = 0$ and $M_\infty + U \in C$.

Note that $E_n\|(M_\infty - M_n) + U\|^2 \le \rho^2(M_n)$ hence $E_n\|(M_\infty - M_n)\|^2 + E_n\|U\|^2 \le \rho^2(M_n)$ and $E_n\|U\|^2 \le \rho^2(M_n) - E_n\|(M_\infty - M_n)\|^2$

$$= \rho^2(M_n) - \sum_{k=0}^{\infty} E_n\|M_{n+k+1} - M_{n+k}\|^2 \le \rho^2(M_n) - E_n\|M_{n+1} - M_n\|^2 \le 2^{-n}.$$

Thus $E\|U\|^2 = 0$, a contradiction.

It follows that $M_\infty \in \mathrm{Ext}(C)$ almost surely and the distribution of M_∞ is a probability μ that verifies the claim of Proposition I.15.

To deduce Theorem I.14, we can again assume that Y is separable. Let T be a dense range compact operator from ℓ_2 into Y. Apply Proposition I.15 to the H_δ-set $T^*(C)$ in ℓ_2 and notice that the extreme points of $T^*(C)$ are exactly the images by T^* of the extreme points of C. Moreover, if μ is a Radon probability measure on $T^*(C)$ representing T^*x then $\tilde{\mu} = \mu T^*$ is a w^*-Radon probability measure on C that represents x.

II. Optimization on non-convex strong w^*-H_δ sets:

In this section, we shall see that the notion of strong H_δ-sets is closely related to the existence of exposing functions for the set in question. The main result of this section is the following:

Theorem II.1: Let C be a bounded subset of a dual Banach space Y* such that C is a strong w^*-H_δ set in $D = \overline{\text{conv}}^*(C)$.

a) If D is w*-metrizable then it is the w*-closed convex hull of its w*-exposed points that are contained in C and the set of functionals in Y that expose D at a point in C is a dense G_δ in Y.

b) If C is norm separable, then D is the w*-closed convex hull of its strongly w*-exposed points that are contained in C and the set of functionals in Y that strongly expose D at a point in C is a dense G_δ in Y.

One should notice that in case b), every strongly w*-exposed point of D is contained in C. This is because strong w^*-H_δ sets are norm-closed, and every non-empty slice of D intersects C. Therefore the norm-closed convex hull of the strongly w*-exposed points of D is not equal to D. For example, if X is a non-reflexive space with separable bidual, B_X is a strong w^*-H_δ in X** by Lemma I.1, and $B_{X^{**}}$ is not the norm-closed convex hull of its strongly w*-exposed points. But this convex hull is weakly dense in B_X, thus norm-dense in B_X, so B_X is the norm-closed convex hull of its strongly exposed points: this is a particular case of Phelps' theorem on RNP spaces. [24]

We shall actually prove a result slightly more general than Theorem II.1, since it applies also if C is not bounded. This formulation is more convenient when applied to the epigraph of a non necessarily bounded function (see Theorem II.11).

Theorem II.1 bis: Let C be a subset of a dual Banach space Y^* that is a strong w^*-H_δ set in $D = \overline{\text{conv}}^*(C)$. Assume that V is a norm-open subset of Y such that $\sup y(C) < \infty$ for every $y \in V$.

If every w^*-compact subset of D is w^*-metrizable (resp: if C is norm-separable) the set of functionals in V that expose (resp: strongly expose) D at a point in C is then a dense G_δ in V.

For each subset L of D we denote by $F(L,D)$ the set of functions in Y that achieve their maximum on D at a point in L, i.e.

$$F(L,D) = \{y \in Y;\ \exists \ell \in L,\ y(\ell) \geq y(d) \quad \text{for every } d \in D\}$$

We shall use the easy fact that $F(L,D)$ is norm-closed when L is w^*-compact.

The actual proof of Theorem II.1 bis starts with Lemma II.8. Before that we shall investigate a notion related to Theorem II.1, and which may have an independent interest: assume now that D is a w^*-compact convex subset of a dual space Y^*. We say that a w^*-compact convex subset K of D is remote in D if for every w^*-compact convex subset G of Y^*, we have

$$D \subseteq \text{conv}(K \cup G) \Rightarrow D \subseteq G.$$

The following lemma is due to Bishop-Phelps (Lemma I.2 [9]).

Lemma II.2: Let f and g be two functionals on a Banach space X such that $\|f\| = \|g\| = 1$. Suppose there exists $\varepsilon > 0$ such that $|g(x)| \leq \varepsilon/2$ whenever $\|x\| \leq 1$ and $f(x) = 0$, then either $\|f-g\|$ or $\|f+g\|$ is less than ε.

Lemma II.3: Let D,K be two w*-compact convex subsets of a dual Banach space Y^* such that $K \subseteq D$ and $K \neq D$. If the set $F(K,D)$ has a non-empty interior, then there exists a w*-compact convex set G such that $D \subseteq \text{conv}(K \cup G)$ but $D \not\subseteq G$ (i.e. K is not remote in D).

Proof: For $z \in Y$, let $M(z,D) = \{x^* \in Y^*;\ z(x^*) = \max z(D)\}$. Let y be in the interior of $F(K,D)$ and let $\varepsilon > 0$ be such that for all z in $B(y,\varepsilon)$, $K \cap M(z,D) \neq \phi$. Note that $y \neq 0$: otherwise, every functional z in Y would attain its maximum on D at a point of K, thus $K = D$.

We can assume without loss of generality $\max y(D) > 0$, $D \subseteq B_{Y^*}$ and that $\|y\| = 1$. Let γ, R be such that $0 < \gamma < \max y(D)$ and $\frac{2}{R} < \inf(\gamma,\varepsilon)$. Consider the w*-compact convex set $G = \{y^* \in Y^*;\ y(y^*) \leq \gamma \text{ and } \|y^*\| \leq R\}$.

To prove that $D \subseteq \text{conv}(K \cup G)$, note that if $c^* \in D$ but $c^* \notin \text{conv}(K \cup G)$, we can find then z in Y, $\|z\| = 1$ and $z(c^*) > \max z(K \cup G)$. We get then: $y(y^*) = 0$ and $\|y^*\| \leq R \Rightarrow y^* \in G \Rightarrow z(y^*) < z(c^*) \leq 1$. Lemma II.2 gives then that either $\|y+z\| \leq \frac{2}{R}$ or $\|y-z\| \leq \frac{2}{R}$. The first case is not possible since $c^* \notin G$ hence $y(c^*) > \gamma$ and $z(c^*) \geq \max z(G) \geq 0$ which implies $(y+z)(c^*) \geq \gamma > \frac{2}{R}$. It follows that $\|y-z\| \leq \frac{2}{R} < \varepsilon$ and $M(z,D) \cap K = \phi$ since $z(c^*) > \max z(K)$: a contradiction.

Finally note that $D \not\subseteq G$ since $\gamma < \max y(D)$ and $\max y(G) \leq \gamma$.

Lemma II.4: Let D,K,G be three w*-compact convex subsets of a dual space Y^* such that: $K \subseteq D$, $D \not\subseteq G$ and $D \subseteq \text{conv}(K \cup G)$. Then for every z in Y such that $\max z(D) > \max z(G)$ and for every $\varepsilon > 0$, there exists $\tau > 0$ such that $S(D,z,\tau) \subseteq K + \varepsilon B_{Y^*}$.

In particular, for every maximizing sequence (x_n^*) for z we have $d(x_n^*,K) \longrightarrow 0$.

Proof: Assume $\alpha = \max z(D) > \max z(G) = \beta$ and let $\tau \leq \frac{(\alpha-\beta)}{d} \varepsilon$ where d is such that $(-K+G) \subseteq dB_{Y^*}$. We can write any c^* in $S(D,z,\tau)$ as $c^* = (1-\theta)k + \theta g$ where $0 \leq \theta \leq 1$, $k \in K$ and $g \in G$. Hence by applying z we get:

$$\alpha - \tau \leq (1-\theta)\alpha + \theta\beta = \alpha - \theta(\alpha-\beta) \text{ and } \theta(\alpha-\beta) \leq \tau \leq \frac{\varepsilon(\alpha-\beta)}{d}.$$

It follows that $0 \leq \theta \leq \frac{\varepsilon}{d}$ and $S(D,z,\tau) \subseteq K + \theta(-K + G) \subseteq K + \varepsilon B_{Y^*}$.

Remark II.5: By combining the above lemmas we get: If $z \in \overset{\circ}{F}(K,D)$ then for any $\varepsilon > 0$, there exists $\tau > 0$ such that $S(D,z,\tau) \subseteq K + \varepsilon B_{Y^*}$. A direct proof is given in Lemma II.8.

Proposition II.6: Let D,K be w*-compact convex subsets of a dual space Y* such that $K \subseteq D$. The following properties are then equivalent:

1) K is remote in D.
2) F(K,D) has an empty interior in Y.

Furthermore, if K is not remote then $d(K,C) = 0$ for any subset C of D such that $D = \overline{\text{conv}}^*(C)$.

Proof: 1) => 2) is the statement of Lemma II.3. For 2) => 1) note that if K is not remote in D then there exists a w*-compact convex set G in Y* such that $D \not\subseteq G$ but $D \subseteq \text{conv}(K \cup G)$. Hence there exists z_0 in Y with $\max z_0(D) > \max z_0(G)$. Note that all z in Y which are close enough to z_0 also verify $\max z(D) > \max z(G)$. By Lemma II.4, $M(z,D) \subseteq K$ for all such z. Hence F(K,D) has a non-empty interior.

If now C is a subset of D such that $D = \overline{\text{conv}}^*(C)$, we have for each z in Y: $\max z(D) = \sup z(C)$. Hence any z in Y has a maximizing sequence in C. This implies, in view of Lemma II.4, that $d(C,K) = 0$ whenever K is not remote in D.

An immediate application of the Baire Category theorem is now the following:

Proposition II.7: Let D be a w*-compact convex subset of a dual space Y*. Let C be a strong w*-H_δ in D such that $D = \overline{\text{conv}}^*(C)$. Then the set of functionals in Y that attain their maximum on D at a point of C contains a dense G_δ in Y.

Proof: Write $D \setminus C = \bigcup_n K_n$ where each K_n is w*-compact convex with $d(K_n, C) > 0$. It follows that each K_n is remote in D and the set $\{y \in Y; y \text{ attains its maximum on D at a point of C}\}$ contains the complement G of $\bigcup_n F(K_n, D)$, and G is a dense G_δ in Y.

Q.E.D.

We begin now the actual proof of Theorem II.1.bis.

Lemma II.8. Let D be a subset of a dual space Y* and K a w*-compact convex subset of D. Suppose $y \in Y$ and $\alpha > 0$ are such that $B(y,\alpha) \subseteq F(K,D)$, then for any $\varepsilon > 0$ we have $S(D,y,\varepsilon) \subset K + \frac{\varepsilon}{\alpha} B_{Y^*}$. In particular $d(K,C) = 0$ for every subset C of D such that $D \subseteq \overline{\text{conv}}^*(\mathbb{C})$.

Proof: Suppose $x^* \notin K + \frac{\varepsilon}{\alpha} B_{Y^*}$; there exists then z in Y, $\|z\| \leq 1$ and $\sup z(K) < z(x^*) - \frac{\varepsilon}{\alpha}$. We claim that $x^* \notin S(D,y,\varepsilon)$. Indeed, if not then $y(x^*) > \max y(D) - \varepsilon$ and $y + \alpha z \in B(y,\alpha) \subseteq F(K,D)$ hence there exists k_0 in K such that $(y+\alpha z)(k_0) \geq (y+\alpha z)(x^*)$. But

$$(y+\alpha z)(x^*) = y(x^*) + \alpha z(x^*) > (y(k_0)-\varepsilon) + \alpha\ (z(k_0)+ \frac{\varepsilon}{\alpha}) = y(k_0) + \alpha z(k_0).$$

A contradiction. The second assertion follows from the fact that $S(D,y,\varepsilon)$ must intersect $\mathbb{C}$ for every $\varepsilon > 0$.

Assume now that C is a subset of Y* such that every w*-compact subset of

$D = \overline{\text{conv}}^*(C)$ is w*-metrizable. We can find in this case a metric δ on D which induces the w*-topology on bounded subsets of D and such that $\delta(d_1,d_2) \leq \|d_1 - d_2\|$ for every $d_1, d_2 \in D$. (Apply Lemma 0 to $D \cap nB_{Y^*}$, for every integer n, in order to get a separable subspace Z of Y which separates the points of D, then set $\delta(d_1,d_2) = \|T^*d_1 - T^*d_2\|$ where T is any compact operator from ℓ^2 into Z, with dense range and $\|T\| \leq 1$.)

Assume further, as in Theorem II.1 bis, that V is a norm-open subset of Y such that $\sup y(C) < \infty$ for every $y \in V$. Remark first that $\sup y(C) = \sup y(D)$. Note also that $S = D \cap \{y^*; y(y^*) > \lambda\}$ is bounded for every $y \in V$ and $\lambda \in \mathbf{R}$; indeed since V is open we can find $\varepsilon > 0$ such that

$$\sup(y+\varepsilon z)(S) \leq \sup(y+\varepsilon z)(D) < \infty$$

for every $z \in B_Y$, and this yields immediately $\sup z(S) < \infty$ for every $z \in Y$. Since S is bounded, we deduce that every $y \in V$ attains its maximum on D.

Recall finally that F(C,D) denotes the (possibly empty) set of functionals in Y which achieve their maximum on D at a point of C.

<u>Lemma II.9</u>. Let C be a subset of a dual space Y* and $D = \overline{\text{conv}}^*(C)$. Assume that V is a norm-open subset of Y such that $\sup y(C) < \infty$ for every $y \in V$, and that F(C,D) contains a dense G_δ-subset of V. Let K be a w*-compact subset of D such that $K \cap C = \emptyset$. If every w*-compact subset of D is w*-metrizable (resp: if C is norm-separable) then for every $\varepsilon > 0$ the set

$$O(K,\varepsilon) = \{y \in V;\ \exists \tau > 0,\ \bar{S}^*(D,y,\tau) \cap K = \emptyset \text{ and } \delta\text{-diam } \bar{S}^*(D,y,\tau) \leq \varepsilon\}$$

(resp the set $O'(K,\varepsilon) = \{y \in V;\ \exists \tau>0,\ \bar{S}^*(D,y,\tau) \cap K = \emptyset$ and $\|\ \|\text{-diam } \bar{S}^*(D,y,\tau) \leq \varepsilon\}$)

is open and dense in V.

In the above statement, δ is a metric on D inducing the w*-topology on every bounded subset of D and such that $\delta(d_1,d_2) \leq \|d_1 - d_2\|$ for all $d_1, d_2 \in D$. In the case where C is bounded, we can apply the lemma with V = Y.

Proof: Since $S(D,y,\tau)$ is bounded for every $y \in V$ it follows easily that $O(K,\varepsilon)$ and $O'(K,\varepsilon)$ are open. To show that they are norm-dense in V, let Ω be any non-empty open subset of V.

Use the w*-metrizability of $D \cap nB_{Y^*}$ for every integer n (resp the norm separability of C) to find a countable family of w*-compact convex subsets (C_k) of D such that:

(a) $C \subseteq \bigcup_k C_k$.

(b) $d(C_k,K) > 0$ for each k.

(c) $\delta\text{-diam}(C_k) \leq \varepsilon/2$ (resp $\|\ \|\text{-diam}(C_k) \leq \varepsilon/2$) for each k.

Note that $\Omega \supset \bigcup_k F(C_k,D) \cap \Omega \supset F(C,D) \cap \Omega$ and the latter contains a dense G_δ set in Ω, hence by Baire's category theorem, there exists k_0 such that $F(C_{k_0},D) \cap \Omega$ has a non-empty interior. If y is in such an interior, we get from Lemma II.8 that for each $\eta > 0$, there exists $\tau > 0$ such that $S(D,y,\tau) \subseteq C_{k_0} + \eta B_{Y^*}$. It is now enough to choose $\eta < \inf\{\varepsilon/2, d(C_{k_0},K)\}$ to obtain that $y \in \Omega \cap O(K,\varepsilon)$ (resp $\Omega \cap O'(K,\varepsilon)$).

Proof of Theorem II.1 bis: Suppose $D \setminus C = \bigcup_n K_n$, where each K_n is w*-compact convex with $d(K_n,C) > 0$. Lemma II.8 gives then that each closed set $F(K_n,D)$ has an empty interior, hence $F(C,D)$ contains a dense G_δ in V (see the proof of Proposition II.7). Lemma II.9 applies and we get that for each N, the set $O_N = O(K_0 \cup K_1 \dots \cup K_N, \frac{1}{N})$ (resp $O'_N = O'(K_0 \cup \dots \cup K_N, \frac{1}{N})$) is a dense open set in V. Note that by the Baire category theorem, the set $O = \bigcap_N O_N$ (resp $O' = \bigcap_N O'_N$) is a dense G_δ in V consisting of the functionals in V that expose (resp strongly expose) D at a point of C.

Now we turn to the problem of minimizing non-linear functions on strong w^*-H_δ sets. Let C be a subset of a dual space Y^* and let ϕ be a $(-\infty,+\infty]$-valued function on C. We shall say that ϕ is a <u>strong w^*-H_δ function</u> if the epigraph of ϕ is a strong w^*-H_δ relative to $C \times \mathbf{R}$: that is $(C \times \mathbf{R})\setminus \mathrm{Epi}(\phi) = \bigcup_n (K_n \cap (C\times\mathbf{R}))$ where each K_n is a w^*-compact convex subset of $Y^* \times \mathbf{R}$ such that $d(K_n, \mathrm{Epi}(\phi)) > 0$. We shall denote by $\mathcal{H}(C)$ the space of such functions on C that are not identically equal to $+\infty$.

<u>Proposition II.10</u>: Let C be a bounded subset of a dual Banach space Y^* and let $D = \overline{\mathrm{conv}}^*(C)$.

a) If D is w^*-metrizable then every weak*-lower semi-continuous function on C belongs to $\mathcal{H}(C)$.

b) Every function in $\mathcal{H}(C)$ is necessarily norm -lower semi-continuous. Conversely, if C is norm separable then every norm-lower semi-continuous function on C is in $\mathcal{H}(C)$.

c) If Y is separable then every w^*-Baire (1) functional in Y^{**} belongs to $\mathcal{H}(D)$, therefore it induces on C a function in $\mathcal{H}(C)$.

d) The supremum of a sequence of functions in $\mathcal{H}(C)$ belongs to $\mathcal{H}(C)$.

<u>Proof</u>: a) follows from Lemma I.1.a) since the epigraph of a weak*-lower semi-continuous function is relatively w^*-closed in $C \times \mathbf{R}$.

For b) it is enough to notice that a strong w^*-H_δ set is necessarily norm-closed. If C is norm separable, the converse follows from Lemma I.(1b).

c) If now f is a weak*-Baire (1) functional in Y^{**}, then by a theorem of

Odell-Rosenthal [23], f is the pointwise limit of a sequence of functionals (f_n) in Y. We may assume without loss of generality that $\|f_n\| \leq 1$ for each n. Note now that the complement in $D \times \mathbf{R}$ of the epigraph of f is $\{(x,\lambda) \in D \times \mathbf{R}; f(x) > \lambda\}$ and is equal to $\bigcup_m K_m$ where each $K_m = \bigcap_{n=m}^{\infty} \{(x,\lambda) \in D \times \mathbf{R};\ f_n(x) \geq \lambda + \frac{1}{m}\}$ is w*-compact and convex. To prove that $d(K_m, \mathrm{Epi}(f)) > 0$, let (x_0,λ_0) be any point in Epi(f) and (x,λ) in K_m. Let $N \geq m$ such that $(f_N - f)(x_0) \leq \frac{1}{2m}$. We have

$$\|x_0 - x\| \geq f_N(x - x_0) \geq \lambda + \frac{1}{m} - f_N(x_0) + f(x_0) - f(x_0)$$
$$\geq \lambda + \frac{1}{m} - \frac{1}{2m} - \lambda_0$$

Hence $\|x_0 - x\| + |\lambda_0 - \lambda| \geq \frac{1}{2m}$ and $d(K_m, \mathrm{Epi}(\phi)) \geq \frac{1}{2m} > 0$.

To prove d) it is enough to notice that a countable intersection of strong w*-H_δ sets is a strong w*-H_δ.

<u>Theorem II.11</u>: Let C be a bounded subset of a dual Banach space Y* such that C is a strong w*-H_δ in $D = \overline{\mathrm{conv}}^*(C)$. If D is w*-metrizable (resp if C is norm separable) then for any bounded below function ϕ in $\mathcal{H}(C)$ and any $\varepsilon > 0$, there exist y in Y, $\|y\| \leq \varepsilon$ and x_0^* in C such that

(i) $(\phi + y)(x^*) > (\phi + y)(x_0^*)$ for all x^* in C, $x^* \neq x_0^*$.

(ii) Every minimizing sequence (x_n^*) for $\phi + y$ in C, w*-converges (resp norm-converges) to x_0^*.

(iii) If ϕ is also concave then x_0^* is an extreme point of C.

<u>Proof</u>: Note that $C \times \mathbf{R}$ is a strong w*-H_δ in $D \times \mathbf{R}$; since $\Phi \in \mathcal{H}(C)$, we get that Epi(Φ) is a strong w*-H_δ in $D \times \mathbf{R}$, therefore also in $\overline{\mathrm{conv}}^*(\mathrm{Epi}(\Phi))$. Since Φ is bounded from below, every function $(y,\alpha) \in Y \times \mathbf{R}$ has a finite supremum on Epi(Φ) when $\alpha < 0$, hence Theorem II.1 bis applies to Epi(Φ), with

$V = \{(y,\alpha);\ y \in Y,\ \alpha < 0\}$. It is thus possible to find for each $\varepsilon > 0$ a functional (y,α) in V such that $\|(y,\alpha) - (0,-1)\| \leq \varepsilon$ and which exposes (respectively strongly exposes) $\mathrm{Epi}(\phi)$ at a point (x_0^*,λ_0). If $0 < \varepsilon < 1/2$, then $-3/2 < \alpha < -1/2$ hence $\lambda_0 = \phi(x_0^*)$. Moreover, for each $(x^*,\lambda) \in \mathrm{Epi}(\phi)$ such that $x^* \neq x_0^*$, we have $y(x_0^*) + \alpha\,\lambda_0 > y(x^*) + \alpha\,\lambda$ hence $\frac{1}{\alpha}\,y(x_0^*) + \phi(x_0^*) < \frac{1}{\alpha}\,y(x^*) + \phi(x^*)$ for each $x^* \neq x_0^*$ in C. Note that if $z = \frac{y}{\alpha}$ then $\|z\| \leq 2\,\varepsilon$ and $\phi + z$ attains its minimum at x_0^* in C.

If (x_n^*) is a minimizing sequence for $\phi + z$ on C then $(x_n^*,\phi(x_n^*))$ is a maximizing sequence for the functional (y,α) that exposes (resp strongly exposes) $\mathrm{Epi}(\phi)$ at (x_0^*,λ_0), hence (x_n^*) w*-converges (resp norm-converges) to x_0^*.

If now ϕ is concave, $\phi + z$ is also concave and x_0^* is then an extreme point of C since it is the only point where $\phi + z$ attains its minimum in C.

<u>Remark II.12</u>: The same proof actually gives under the hypothesis of Theorem II.11 that for any bounded below function ϕ in $\mathcal{H}(C)$, the set $\{y \in Y\ ;\ \phi + y$ exposes (respectively strongly exposes) C from below$\}$ is a dense G_δ in Y.

The following is an extension of a result due to Haydon [10] in the case of weak*-compact convex sets.

<u>Corollary II.13</u>: Let Y be a Banach space not containing an isomorphic copy of ℓ_1. Let C be a convex bounded subset of Y* such that $\bar{C}^*$ is w*-metrizable. If C is a strong w*-H_δ in $\bar{C}^*$ then C is the norm closed convex hull of its exposed points.

<u>Proof</u>: By Lemma (0), we can assume without loss of generality that Y is separable. Since C is a strong w*-H_δ set in $\bar{C}^*$, it is automatically norm closed. Let F be the norm closed convex hull of the exposed points of C. If

$F \neq C$, there exists $\alpha > 0$ and f in Y^{**} such that $F \cap \{x^* \in C; f(x^*) \geq M(f,C) - \alpha\} = \emptyset$. By a theorem of Odell-Rosenthal [23] f is weak*-Baire 1, hence $f \in \mathcal{H}(C)$. Apply Theorem II.11 to get for each $\varepsilon > 0$, a functional y in Y and x_0^* in C such that $\|y\| \leq \varepsilon$ and $(f+y)(x^*) < (f+y)(x_0^*)$ for all x^* in C with $x^* \neq x_0^*$. Hence x_0^* is an exposed point for C. Moreover, if $\varepsilon < \alpha/2K$ where $K = \sup\{\|x^*\|; x^* \in C\}$ then $f(x_0^*) \geq M(f,C) - \alpha$ and $x_0^* \notin F$, a contradiction.

We now give the following converse to Theorem II.1.

Theorem II.14: Let C be a bounded convex subset of a dual space Y^*. If C is a w^*-G_δ in a w^*-metrizable $D = \bar{C}^*$ (resp if C is norm separable) then the following conditions are equivalent:

1) C is a strong w^*-H_δ in D

2) For every bounded below convex and w^*-lower semi-continuous (resp norm-lower semi-continuous) extended-real-valued function ϕ on C with $\phi \not\equiv +\infty$, the set $F(\phi,C) = \{y \in Y; \phi+y$ attains its minimum in $C\}$ contains a dense G_δ in Y.

Proof: Note that 1) => 2) is a weak form of Theorem II.11 and Remark II.12. For 2) => 1) note first that the hypothesis implies that C is norm closed. Indeed, let x_0 be in the norm closure of C. Let $\phi(x) = \|x - x_0\|$ for each x in C and take $0 < \varepsilon < 1$. There exists then y in Y, $\|y\| \leq \varepsilon$ such that $\phi + y$ attains its minimum at a point c_0 in C. That is

$$\|c - x_0\| \geq \| c_0 - x_0\| - y(c - c_0) \text{ for all c in C.}$$

Since $\|y\| \leq \varepsilon$ we get $\|c_0 - x_0\| \leq \|c - x_0\| + \varepsilon \| c - c_0\|$ for all c in C. Let now (c_n) be a sequence in C converging to x_0. We have $\|c_0 - x_0\| \leq \lim_n(\|c_n - x_0\| + \varepsilon \|c_n - c_0\|) = \varepsilon\|x_0 - c_0\|$. Hence $x_0 = c_0$ belongs to C.

Suppose first that C is a w^*-G_δ in D and that the latter is w*-metrizable. Let δ be a metric on D that induces the w*-topology and let Δ be a w*-lower semi-continuous metric on D that induces the w*-topology on C while making it complete. We shall prove that C is w*-Δ-dentable. For that let L be any relatively w*-closed convex subset of C. By Lemma I.1.a, $\bar{L}^* \setminus L = \bigcup_n K_n$ where each K_n is w*-compact. Let now ϕ be the convex and w*-l.s cont function on D defined by $\phi(x) = \begin{cases} 0 & \text{if } x \in \bar{L}^* \\ +\infty & \text{if } x \in D \setminus \bar{L}^* \end{cases}$

Suppose $y \in F(\phi, C)$, that is there exists x_0 in C such that

$$\phi(x_0) + y(x_0) \leq \phi(x) + y(x) \text{ for all } x \text{ in } C.$$

If $x \in L$, we have $\phi(x_0) + y(x_0) \leq y(x)$. Hence $\phi(x_0) = 0$ and $x_0 \in C \cap \bar{L}^* = L$. It follows that $F(\phi, C)$ contains a dense G_δ in Y contained in $F(L, \bar{L}^*)$. Lemma II.9 applies to L and we get that for each N, the open set $O_N = \{y \in Y;\ \exists \tau > 0,\ \overline{S(\bar{L}^*, y, \tau)}^* \cap (K_0 \cup \ldots \cup K_N) = \emptyset \text{ and } \delta\text{-diam } \overline{S(\bar{L}^*, y, \tau)}^* \leq \frac{1}{N}\}$ is dense in Y. The set $O = \bigcap_N O_N$ is then a dense G_δ in Y consisting of the functionals that expose $\bar{L}^*$ at a point of L. It follows that L is contained in the w*-closed convex hull of its w*-exposed points. These points have w*-slices of arbitrarily small Δ-diameter. It follows that C is w*-Δ-dentable and Proposition I.2 gives that C is a w*-H_δ in D.

In the case where C is norm separable, Hypothesis 2) gives as above that for every norm closed convex subset L of C, the set $F(L,\overline{L}^{*})$ contains a dense G_δ in Y. Lemma II.9 then gives that for each N, the set $O''_N = \{y \in Y;\ \exists \tau > 0,\ \|\ \|\text{-diam}\ \overline{S(\overline{L}^{*},y,\tau)}^{*} \le \frac{1}{N}\}$ is dense in Y. The set $O'' = \bigcap_N O''_N$ is then a dense G_δ in Y. Since L is norm closed, the functions in O" strongly expose L; hence C is w*-dentable and Proposition I.2 applied to the metric induced by the norm gives that C is a w*-H_δ in D.

To conclude that C is a strong w*-H_δ in D we need Corollary III.4 whose proof is delayed until the next section.

III. WHEN H_δ- SETS ARE STRONG H_δ-SETS:

We shall see in section VI that there exists a separable closed convex w^*-H_δ set which is not a strong w^*-H_δ set. In this section we exhibit few cases where the two notions coincide. We shall say that a subset C of a dual space Y* has the *-Brondsted-Rockafellar property if for any bounded convex and w*-lower semi-continuous function ϕ on C the set $F(\phi,C)=\{y \in Y:\phi + y$ attains its minimum on C$\}$ is dense in Y. Note that $F(\phi,C)$ is not assumed to be a G_δ and that a nice application of the Bishop-Phelps theorem [9] isolated by Brondsted-Rockafellar [5] gives that any closed convex bounded subset of a Banach space X, has the *-Brondsted-Rockafellar property once considered as a subset of X**.

We first isolate two lemmas that will be of frequent use in the rest of this paper.

Lemma III.1: Let C and K be two non-empty bounded subsets of a dual Banach space Y* such that $C \cap K = \phi$ and K is w*-metrizable. Assume K is w*-compact (resp w*-compact and convex) and that for any non-empty w*-compact (resp w*-compact convex) subset L of K there exists a w*-open set (resp a w*-open half-space) V such that $L \cap V \neq \emptyset$ and $d(L \cap \bar{V}^*, C) > 0$. Then K can be written as a countable union of w*-compact (resp w*-compact and convex) subsets (K_n) such that $d(K_n,C) > 0$ for each n.

Proof: By transfinite induction, we define a decreasing family (L_α) of w*-compact (resp w*-compact and convex) subsets of K in the following manner:

(a) $L_0 = K$.

(b) If $\alpha = \beta + 1$ and L_β non-empty, use the hypothesis to find a w*-open set (resp a w*-open half-space) V_β such that $L_\beta \cap V_\beta \neq \emptyset$ and $d(L_\beta \cap \bar{V}_\beta^*, C) > 0$. Then set $L_\alpha = L_\beta \setminus V_\beta$.

(c) If α is a limit ordinal set $L_\alpha = \bigcap_{\beta<\alpha} L_\beta$.

Since K is w*-metrizable, there exists $\gamma < \Omega$ (the first uncountable ordinal) such that $L_\gamma = \emptyset$. It is clear that $K = \bigcup_{\alpha<\gamma} L_\alpha \cap \bar{V}_\alpha^*$ and that each $K_\alpha = L_\alpha \cap \bar{V}_\alpha^*$ is a strictly positive distance away from C.

<u>Lemma III.2</u>: Let C and K be two non-empty bounded subsets of a dual Banach space Y* such that $C \cap K = \emptyset$ and C is norm separable. Assume K is w*-compact (resp w*-compact and convex) and that for any non-empty w*-compact (resp w*-compact and convex) subset L of K, there exists a w*-open set (resp a w*-open half-space) V and an x in C such that $d(C, L \cap \bar{V}^*) \geq \frac{1}{2} d(x,L)$ and $d(x, L \setminus V) \geq \frac{3}{2} d(x,L)$. Then K can be written as a countable union of w*-compact (resp w*-compact and convex) subsets (K_n) such that $d(K_n, C) > 0$.

<u>Proof</u>: By transfinite induction, we define a decreasing family (L_α) of w*-compact (resp w*-compact and convex) subsets of K, and a family (x_α) of points in C, in the following manner:

(a) $L_0 = K$.

(b) If $\alpha = \beta + 1$ and L_β is non-empty, use the assumption to find x_β in C and a w*-open set (resp a w*-open half-space) V_β such that

(i) $d(C, L_\beta \cap \bar{V}_\beta^*) \geq \frac{1}{2} d(x_\beta, L_\beta)$ and

(ii) $d(x_\beta, L_\beta \setminus V_\beta) \geq \frac{3}{2} d(x_\beta, L_\beta)$

and let $L_\alpha = L_\beta \setminus V_\beta$.

(c) If α is a limit ordinal, set $L_\alpha = \bigcap_{\beta<\alpha} L_\beta$.

Note that the family $(\phi_\alpha) = \{d(\cdot, L_\alpha);\ L_\alpha \text{ non-empty}\}$ is a bounded and increasing net of continuous functions on C. Since the latter is separable, there exists $\gamma < \Omega$ such that $\phi_\gamma = \phi_\eta$ for all $\eta \geq \gamma$. This implies that $L_\eta = \emptyset$ for all $\eta > \gamma$. Indeed, if not there exists $x_\eta \in C$, $L_{\eta+1} \subseteq L_\eta$ such that $\phi_{\eta+1}(x_\eta) \geq \frac{3}{2}\phi_\eta(x_\eta)$ which contradicts the stationarity of (ϕ_α) after γ. It follows as in Lemma III.1 that $K = \bigcup_{\alpha<\gamma} L_\alpha \cap \bar{V}^*_\alpha$.

<u>Theorem III.3</u>: Let C be a subset of a dual space Y* that verifies the *-Brondsted-Rockafellar property then:

(a) Any w*-metrizable w*-compact and convex subset K of Y* that is disjoint from C, can be written as a countable union of w*-compact and convex subsets (K_n) that are a strictly positive distance away from C.

(b) If C is norm separable then the same holds for any w*-compact convex set disjoint from C.

<u>Proof</u>: In view of the above lemmas, it is enough to show that for any non-empty w*-compact convex subset L of K, there exists x_0 in C and a w*-open half-space H such that

(i) $d((C, L \cap \bar{H}^*) \geq \frac{1}{2} d(x_0, L)$ and (ii) $d(x_0, L \setminus H) \geq \frac{3}{2} d(x_0, L)$.

To do so, apply the *-Brondsted-Rockafellar property to the bounded convex and w*-lower semi-continuous function $\phi(x) = d(x,L)$ and $\varepsilon = 1/4$ to get x_0 in C, y in Y with $\|y\| \leq 1/4$ such that:

$$(*) \qquad d(x,L) \geq d(x_0,L) - y(x-x_0) \text{ for all } x \text{ in } C.$$

Let $r_0 = d(x_0,L)$ and $H = \{x \in Y^*;\ y(x - x_0) < \frac{3}{8} r_0\}$. To prove (i) note that (*) implies that $d(x,L) \geq \frac{r_0}{2}$ for all x in C such that $y(x-x_0) \leq \frac{r_0}{2}$. On the other hand if $y(x - x_0) > \frac{r_0}{2}$, take any k in $\bar{H}^*$ and we get

$$\|x - k\| \geq 4y[(x - x_0) - (k - x_0)] \geq 4[\frac{1}{2} r_0 - \frac{3}{8} r_0] = \frac{1}{2} r_0.$$

Hence $d(C, L \cap \overline{H}^*) \geq \frac{1}{2} r_0$. For (ii) take any k in $L \setminus H$; then $\|x_0 - k\| \geq 4y(k - x_0) \geq 4.\frac{3}{8} r_0 = \frac{3}{2} r_0$ hence $d(x_0, L \setminus H) \geq \frac{3}{2} r_0$. Note that (ii) implies in particular that $L \cap H \neq \emptyset$.

Now we can deduce the following

Corollary III.4: Let C be a bounded subset of a dual Banach space Y* such that C is a w*-H_δ in $D = \overline{\text{conv}}^*(C)$. If either C is norm separable or D is w*-metrizable, then the following conditions are equivalent:

1) C is a strong w*-H_δ in D.

2) C verifies the *-Brondsted-Rockafellar property.

Proof: Note that 1) => 2) is a weak form of Theorem II.11. For 2) => 1) write first $D \setminus C = \bigcup_n K_n$ where each K_n is w*-compact and convex, then apply Theorem III.3 to split each K_n into a countable union of w*-compact and convex sets $(K_{nm})_m$ such that $d(K_{nm}, C) > 0$.

Corollary III.5: Let C be a bounded closed convex subset of a Banach space X such that C is a w*-H_δ in its w*-closure D in X**. If C is norm separable then C is a strong w*-H_δ in D.

Proof: It is enough to notice that C as a subset of X** has the *-Brondsted Rockafellar property.

Corollary III.6: Let C be a separable closed convex bounded subset of a Banach space X. The following conditions are then equivalent:

1) C is a norm H_δ-set in X.

2) C is a strong norm H_δ-set in X.

3) C is a countable intersection of closed half-spaces.

Proof: 1) => 2) Assume $X \setminus C = \bigcup_n F_n$ where each F_n is convex closed and bounded. Let K_n be the w*-closure of F_n in X**. Note that $K_n \cap C = \emptyset$. Since C has the *-Brondsted-Rockafellar property in X**, we can use Theorem III.3 to split each K_n into a countable union of w*-compact and convex subsets $(K_{n,m})_m$ in X** such $d(K_{n,m},C) > 0$ for each m.

Note now that $X \setminus C = \bigcup_n \bigcup_m F_{nm}$ where each $F_{nm} = F_n \cap K_{nm}$ is closed convex bounded in X and $d(F_{nm},C) > 0$.

The rest follows from the Hahn-Banach theorem.

Corollary III.7: Let Z be a separable subspace of a Banach space X. The following conditions are equivalent:

1) Z is a norm H_δ-subspace of X.

2) B_Z is a strong norm H_δ-subset of B_X.

3) The orthogonal of Z in X* is w*-separable.

Proof: 1) <=> 2) follows from Corollary III.6 and the fact that if $B_X \setminus B_Z = \bigcup_n K_n$ then $X \setminus Z = \bigcup_m \bigcup_n m K_n$.

For 3) => 1) Let (f_n) be a w*-dense sequence in $Z^\perp$ that is $Z = \{x \in X; f_n(x) = 0 \text{ for all } n\}$. It is easy to see that for each n, Ker f_n is an H_δ in X.

2) => 3) Let $X \setminus Z = \bigcup_n F_n$ with $d(F_n,Z) > 0$ and let $\pi : X \to X/Z$ be the quotient map. Note that $0 \notin \overline{\pi(F_n)}$ for each n, hence by the Hahn-Banach theorem, there exists f_n in $(X/Z)^* = Z^\perp$ such that $f_n > 0$ on F_n. It follows that $Z = \{x \in X; f_n(x) = 0 \text{ for all } n\}$.

Proposition III.8. Let Y be a Banach space not containing an isomorphic copy of ℓ_1. Let C be a separable closed convex bounded subset of Y* such that $\bar{C}^*$ is w*-metrizable. Then C is a norm H_δ-set in $\bar{C}^*$ if and only if it is a strong w*-H_δ subset of $\bar{C}^*$.

Proof. By lemma (0), we can assume without loss of generality that Y is separable. If C is a norm H_δ-set in $\bar{C}^*$, use Corollary III.6 to write it as a countable intersection of closed half-spaces. Note now that if $H = \{f \leq \alpha\}$ is a half-space with f in Y**, we can find by a Theorem of Odell-Rosenthal [23] a sequence (f_n) in Y that converges pointwise to f on the elements of Y*. It follows that $\{f > \alpha\} = \bigcup_m \bigcap_{n=m}^{\infty} \{f_n \geq \alpha + \frac{1}{m}\}$ and H is then a strong w*-H_δ subset of Y*. The same obviously holds for C.

Remark III.9. We shall see in the next section that these sets are exactly the separable Radon-Nikodym subsets of Y*, provided ℓ_1 does not embed in Y.

Now, we deal with the case of w*-H_δ subspaces. The remainder of this section is devoted to the proof of the following:

Theorem III.10. Let X be a separable subspace of the dual of a separable Banach space Y such that B_X is w*-dense in B_{Y^*}. Then B_X is a w*-G_δ (resp w*-H_δ) in B_{Y^*} if and only if it is a strong w*-G_δ (resp strong w*-H_δ) in B_{Y^*}.

The proof will be broken into several lemmas. We shall need the following terminology: If X is a subspace of Y* and L is a w*-compact subset of Y* which is disjoint from X, we shall say that L is ρ-bad with respect to X for some $\rho > 0$ if

(i) For each $\varepsilon > 0$, the set $L_\varepsilon = \{\ell \in L; d(\ell,X) \leq \varepsilon\}$ is w*-dense in L.

(ii) The set $L^{\rho} = \{\ell \in L; d(\ell,X) \geq \rho\}$ is also w*-dense in L.

For an example of a w*-compact subset of ℓ_∞ that is ρ-bad with respect to c_0, we refer the reader to VI.9.

<u>Lemma III.11.</u> Let X be a subspace of the dual of a separable Banach space Y such that B_X is w*-dense in B_{Y*}. Let L be a w*-compact subset of θB_{Y*} $(\theta < 1)$ which is ρ-bad with respect to X for some $\rho > 0$. Let K be a w*-compact subset of B_{Y*} which is disjoint from X. Then for each ε such that $0 < \varepsilon < 1 - \theta$ and each w*-open subset V of B_{Y*} with $V \cap L \neq \emptyset$, there exist $\ell \in L$, $x \in X$ such that:

(i) $\|x\| \leq \varepsilon$

(ii) $\ell + x \in V$

(iii) $\ell + x \notin K$

(iv) $d(\ell + x, X) \geq \rho - \varepsilon$.

<u>Proof:</u> Choose ℓ_1 in L such that $\ell_1 \in V$ and $d(\ell_1, X) < \frac{\varepsilon}{2}$. Consider x_1 in X and (y_n) in X such that $\ell_1 = \text{w*-}\lim_n (x_1 + y_n)$ with $\|y_n\| \leq \frac{\varepsilon}{2}$. Note that $\|x_1\| \leq \theta + \frac{\varepsilon}{2}$, hence $x_1 + y_n \in B_X$. Choose now n_0 such that $x_1 + y_{n_0} \in V$. Since $K \cap X = \emptyset$ we have that $x_1 + y_{n_0} \notin K$. On the other hand, $\text{w*-}\lim_n(\ell_1 - y_n + y_{n_0}) = x_1 + y_{n_0}$ and $\|\ell_1 - y_n + y_{n_0}\| \leq 1$. Hence for a large enough n_1 we have $(\ell_1 - y_{n_1} + y_{n_0}) \in V$ and $(\ell_1 - y_{n_1} + y_{n_0}) \notin K$.

Since L is ρ-bad, choose (ℓ_m) in L, $d(\ell_m, X) \geq \rho$ such that $\ell_1 = \text{w*-}\lim(\ell_m)$. That is for a large enough m_0 we have $(\ell_{m_0} - y_{n_1} + y_{n_0}) \in V$ and $(\ell_{m_0} - y_{n_1} + y_{n_0}) \notin K$. Take now $\ell = \ell_{m_0}$ and $x = (-y_{n_1} + y_{n_0})$. They clearly verify the claimed properties.

<u>Lemma III.12:</u> Let X be a separable subspace of the dual of a separable Banach space Y such that B_X is a w* - G_δ, w*-dense in B_{Y*}. Then for each $\theta < 1$, θB_{Y*}

contains no ρ-bad w*-compact sets with respect to X for any $\rho > 0$.

Proof: Write $B_{Y*} \setminus B_X = \bigcup_n K_n$ where (K_n) is an increasing sequence of w*-compact sets. Let Δ be a distance defining the w*-topology on B_{Y*} and let (z_n) be a dense sequence in B_X. Suppose L is a weak*-compact subset of θB_{Y*} $(\theta < 1)$ which is ρ-bad with respect to X for some $\rho > 0$.

Let ℓ_o be any point in L and let $V_o = B_\Delta(\ell_o, 1)$ and $0 < \varepsilon_o < \inf(1-\theta, \rho/2)$. Use Lemma III.11 to obtain $x_o \in X$, $\|x_o\| \leq \varepsilon_o$ and $\ell'_o \in L$ such that $\ell'_o + x_o \in V_o, \ell'_o + x_o \notin K_o$ and $d(\ell'_o + x_o, X) \geq \rho - \varepsilon_o > \frac{\rho}{2}$.

Set now $L_1 = L + x_o$, $\theta_1 = \theta + \varepsilon_o < 1$ and $\ell_1 = \ell'_o + x_o$.

Note that L_1 is also ρ-bad with respect to X and $L_1 \subseteq \theta_1 B_{Y*}$. Let V'_1 be a w*-open subset of V_o containing ℓ_1 such that $\bar{V}_1'^{*} \cap (K_o \cup \bar{B}^*(z_o, \frac{\rho}{2})) = \emptyset$. Set $V_1 = V'_1 \cap B_\Delta(\ell_1, 1/2)$ and $\varepsilon_1 < \inf(1 - \theta_1, \rho/2)$ and apply again Lemma III.11 to obtain $x_1 \in X$, $\|x_1\| \leq \varepsilon_1$, $\ell'_1 \in L_1$, $\ell'_1 + x_1 \in V_1$, $\ell'_1 + x_1 \notin K_1$ and $d(\ell'_1 + x_1, X) \geq \rho - \varepsilon_1 > \frac{\rho}{2}$.

By induction, we get a decreasing sequence (V_n) of w*-open subsets of B_{Y*} and a sequence (ℓ_n) of vectors in B_{Y*} such that:

(i) $\ell_n \in V_n$ for each n.

(ii) $\operatorname{diam}_\Delta(\bar{V}_n^*) \leq 2^{-n}$.

(iii) $\bar{V}_n^* \cap (K_{n-1} \quad (\bigcup_{j=0}^{n-1} \bar{B}^*(z_j, \rho/2)) = \phi$.

It follows that the w*-limit ℓ_∞ of (ℓ_n) can neither be in B_X nor in any of the K_n's which is obviously a contradiction since $\ell_\infty \in B_{Y*}$.

Lemma III.13. Let X be a separable subspace of the dual of a separable Banach space Y such that B_X is a w*-G_δ, w*-dense in B_{Y*}. Let L be a subset of Y* which is disjoint from X. Then:

(i) If L is w*-compact, there exists a w*-open set V such that $L \cap V \neq \phi$ and $d(L \cap \bar{V}^*, X) > 0$.

(ii) If L is w*-compact and convex, there exists a w*-open half space V such that $L \cap V \neq \phi$ and $d(L \cap \bar{V}^*, X) > 0$.

Proof: We claim first that there exists $\varepsilon > 0$ such that the set $L_\varepsilon = \{\ell \in L; d(\ell,X) \leq \varepsilon\}$ is not w*-dense in L. Indeed, suppose not. We can assume without loss that $L \subseteq \frac{1}{2} B_{Y^*}$. Note now that $L = \bigcup_n L^n$ where $L^n = \{\ell; d(\ell,X) \geq \frac{1}{n}\}$ since $L \cap X = \phi$. It follows that there exists n_0 such that $\bar{L^{n_0}}^*$ has a non-empty interior V_0 in the w*-topology relative to L. It follows that $\bar{V}_0^*$ is a $\frac{1}{n_0}$-bad set with respect to X which clearly contradicts Lemma III.12.

In case (i) we take V to be a w*-open subset of Y* such that $V \cap L = L \setminus \bar{L}_\varepsilon^*$ which is non-empty.

In case (ii), note that $\bar{L}_\varepsilon^*$ is also convex hence any w*-open half space V that separates any point ℓ in $L \setminus \bar{L}_\varepsilon^*$ from $\bar{L}_\varepsilon^*$ will do the job.

To prove Theorem III.10, it is enough to write $B_{Y^*} \setminus B_X = \bigcup_n K_n$ were each K_n is w*-compact (resp w*-compact and convex) and then to apply Lemma III.13 and Lemma III.1 to B_X and each K_n to obtain the claimed result.

IV. ON THE REPRESENTATION OF RADON-NIKODYM SETS AS STRONG w^*-H_δ SETS:

Recall first that a bounded subset C of a Banach space X is said to have the point of weak to norm continuity property (P.C.P) (resp the Radon-Nikodym property (R.N.P)) if for every $\varepsilon > 0$ and every non-empty subset F of C, there exists a weakly open set (resp an open half-space) V such that $V \cap F \neq \emptyset$ and $\operatorname{diam}(V \cap F) \leq \varepsilon$. We shall say that a Banach space has the P.C.P. (resp the R.N.P) if its unit ball does.

In the first part of this section we shall give the representation of a Radon-Nikodym set as a w^*-H_δ set in a dual space and some of its applications. The representation as a strong w^*-H_δ set is more involved and its proof will be delayed until the end of the section.

Lemma IV.1: Let C be a separable bounded closed subset of a Banach space X. If C is a P.C.P. set (resp a convex R.N.P. set) then there exists a sequence (K_n) of w^*-compact (resp w^*-compact and convex) subsets of X^{**} and a sequence (O_n) of weak*-open sets (resp of countable unions of w^*-open half spaces) of X^{**} that are determined by a countable set D of functionals in X^* such that:

If Y is any subspace of X^* containing D and a norming set for X and if T is the canonical embedding of Y into X^* then:

$$T^*(C) = \bigcap_n \left(T^*(K_n) \cup T^*(O_n) \right)$$

Proof: Given $\varepsilon > 0$, we construct by transfinite induction a decreasing family of closed (resp closed and convex) subsets (C_α) of C in the following manner:

(i) $C_0 = C$.

(ii) If $\alpha = \beta + 1$ and $C_\beta \neq \emptyset$, use the P.C.P. (resp the R.N.P) to find an elementary w^*-open set (resp a w^*-open half-space) V_β such that

$V_\beta \cap C_\beta \neq \emptyset$ and $\mathrm{diam}(C_\beta \cap V_\beta) \leq \varepsilon$. Then set $C_\alpha = C_\beta \setminus V_\beta$.

(iii) If α is a limit ordinal, let $C_\alpha = \bigcap_{\beta<\alpha} C_\beta$.

Since C is separable, there is $\gamma_\varepsilon < \Omega$ (the first uncountable ordinal) such that $C_{\gamma_\varepsilon} = \emptyset$. Let now K_α be the weak*-closure of C_α in X^{**}.

Note that $C \subseteq K_\alpha \cup (\bigcup_{\beta<\alpha} V_\beta)$ for each $\alpha \leq \gamma_\varepsilon$. On the other hand, if $x^{**} \in \bigcap_{\alpha\leq\gamma_\varepsilon} (K_\alpha \cup (\bigcup_{\beta<\alpha} V_\beta))$, we let α_0 be the first ordinal less than γ_ε such that $x^{**} \notin K_{\alpha_0}$. We have then $x^{**} \in \bigcup_{\beta<\alpha_0} V_\beta$. Hence there exists $\beta < \alpha_0$ such that $x^{**} \in K_\beta \cap V_\beta$. Let now (x_j) be a net in C_β such that $w^*\text{-}\lim_j(x_j) = x^{**}$. If j is large enough we get that $x_i \in V_\beta$ for all $i \geq j$. It follows that $\|x_j - x^{**}\| \leq \underline{\lim}_i \|x_j - x_i\| \leq \varepsilon$. Hence

$$C \subseteq \bigcap_{\alpha\leq\gamma_\varepsilon} (K_\alpha \cup (\bigcup_{\beta<\alpha} V_\beta)) \subseteq C + \varepsilon B_{X^{**}}.$$

Let now D_ε be the countable set of functionals determining the w*-open sets $\{V_\beta;\ \beta \leq \gamma_\varepsilon\}$ and let Y be any closed subspace of X^* containing D_ε and a norming set for X. Let $T:Y \longrightarrow X^*$ be the canonical embedding. Note that $T^*:X^{**} \longrightarrow Y^*$ is a quotient map whose restriction to X is an isometry.

If x is in X, $\alpha > 0$ and G is a finite subset of D_ε (resp $G = \{x^*\}$ is a singleton in D_ε) then the w*-open set $V = V(x,G,\alpha) =$

$\{x^{**} \in X^{**}; \sup_{x^*\in G} |(x^{**} - x)(x^*)| < \alpha\}$ (resp the w*-open half space $H = H(x^*,\alpha) = \{x^{**} \in X^{**};\ x^{**}(x^*) < \alpha\}$) in X^{**} is the inverse image by T^* of the corresponding w*-open set $\tilde V = \tilde V(Tx,G,\alpha)$ (resp $\tilde H = \tilde H(x^*,\alpha)$) in Y^*. Note now that $T^*(\bigcap_{\alpha\leq\gamma_\varepsilon} (K_\alpha \cup (\bigcup_{\beta<\alpha} V_\beta))) \subseteq \bigcap_{\alpha\leq\gamma_\varepsilon} (T^*(K_\alpha) \cup (\bigcup_{\beta<\alpha} (T^*(V_\beta))) = \bigcap_{\alpha\leq\gamma_\varepsilon} (T^*(K_\alpha) \cup (\bigcup_{\beta<\alpha} \tilde V_\beta))$.

On the other hand, if y^* belongs to the right hand side, there exists a directed upward subset M of $\{\alpha \leq \gamma_\varepsilon\}$ such that

$$y^* \in (\bigcap_{m\in M} T^*(K_m)) \cup (\bigcap_{m\notin M} \tilde O_m) \text{ where } \tilde O_\alpha = \bigcup_{\beta<\alpha} \tilde V_\beta.$$

Since the $(K_m)_m$ is a decreasing net of w*-compact sets, we have

$T^*(\bigcap_m K_m) = \bigcap_m T^*(K_m)$.

Moreover, $\bigcap_{m\notin M} \tilde{O}_m = T^*(T^{*-1}(\bigcap_{m\notin M} \tilde{O}_m)) = T^*(\bigcap_{m\notin M} T^{*-1}(\tilde{O}_m))$. Hence

$$y^* \in T^*(\bigcap_{m\in M} K_m) \cup T^*(\bigcap_{m\notin M} T^{*-1}(\tilde{O}_m)) \subset T^*(\bigcap_{\alpha\le\gamma_\varepsilon} (K_\alpha \cup O_\alpha)) = T^*(\bigcap_{\alpha\le\gamma_\varepsilon} (K_\alpha \cup (\bigcup_{\beta<\alpha} V_\beta))).$$

It follows that

$$T^*(C) \subseteq T^*(\bigcap_{\alpha\le\gamma_\varepsilon} (K_\alpha \cup (\bigcup_{\beta<\alpha} V_\beta))) = \bigcap_{\alpha\le\gamma_\varepsilon} (T^*(K_\alpha) \cup (\bigcup_{\beta<\alpha} T^*(V_\beta))) \subseteq T^*(C) + \varepsilon B_{Y^*}$$

By repeating the construction for each $\varepsilon = \frac{1}{n}$ and by taking any subspace Y of X* containing a norming set for X and the set $D = \bigcup_n D_{1/n}$, we get that

$$T^*(C) = \bigcap_n \bigcap_{\alpha\le\gamma_n} (T^*(K_{\alpha,n}) \cup (\bigcup_{\beta<\alpha} T^*(V_{\beta,n}))).$$

<u>Proposition IV.2</u>: Let X be a Banach space with a countable norming set and let C be a separable closed bounded subset of X. If C is a P.C.P. set (resp a convex R.N.P. set) then there exists a separable Banach space Y and an isometric embedding $T: X \longrightarrow Y^*$ such that T(C) is a w^*-G_δ (resp a w^*-H_δ) set in Y* and $T(B_X)$ is w*-dense in B_{Y^*}.

<u>Proof</u>: Since X has a countable norming set, Lemma IV.1 gives then a separable subspace Y of X* such that if $S: Y \longrightarrow X^*$ is the canonical embedding then $S^*: X^{**} \longrightarrow Y^*$ is a quotient map and the restriction T of S* to X is an isometric embedding. Moreover $T(C) = \bigcap_n (S^*(K_n) \cup S^*(O_n))$ where each K_n is w*-compact (resp w*-compact and convex) and each $S^*(O_n)$ is a w*-open set (resp a countable union of w*-open half-spaces) in Y*. Since B_{Y^*} is w*-metrizable, each $S^*(K_n)$ is a countable intersection of w*-open half-spaces. Hence S*(C) is a w^*-G_δ (resp a w^*-H_δ) in Y*. Since B_X is w*-dense in $B_{X^{**}}$, it follows that $T(B_X)$ is w*-dense in $B_{Y^*} = S^*(B_{X^{**}})$.

Corollary IV.3: Let C be a closed bounded subset of a Banach space X such that its weak*-closure in X** is w*-metrizable.

a) If C is a P.C.P. set then C is a w*-G_δ in $\bar{C}^*$.

b) If C is a convex R.N.P. set then C is a strong w*-H_δ in $\bar{C}^*$.

Proof: a) Use Lemma IV.1 with Y = X* to write $C = \bigcap_n (K_n \cup O_n)$ where each K_n is w*-compact in $\bar{C}^*$ and each O_n is w*-open in X**. Since $\bar{C}^*$ is w*-metrizable, each K_n and hence C is a w*-G_δ in $\bar{C}^*$.

For b) note that each K_n is then a w*-H_δ in $\bar{C}^*$, hence C is also a w*-H_δ in $\bar{C}^*$. Corollary III.5 then implies that C is a strong w*-H_δ in $\bar{C}^*$.

Corollary IV.4: Let C be a separable closed convex bounded subset of a dual Banach space Y* such that $\bar{C}^*$ is w*-metrizable. If C has R.N.P. then it is a strong norm H_δ-set in $\bar{C}^*$.

Proof: By Lemma 0, we can assume without loss of generality that Y is separable. Apply IV.2 to C and the space X = Y* to obtain a separable Banach space Z such that Y* can be identified to a subspace of Z* in such a way that $Z^* \setminus C = \bigcup_n K_n$ where each K_n is convex and w*-compact in Z*. Note now that $\bar{C}^* \setminus C = \bigcup_n K_n \cap \bar{C}^*$ hence C is a norm H_δ-set in $\bar{C}^*$ (the w*-closure of C in Y*). The proof of Corollary III.6 gives then that C is a strong norm H_δ in $\bar{C}^*$.

Theorem IV.5: Let Y be a Banach space not containing an isomorphic copy of ℓ_1. Let C be a separable closed convex bounded subset of Y* such that $\bar{C}^*$ is w*-metrizable. The following properties are then equivalent:

1) C has R.N.P.

2) C is a strong w*-H_δ set in $\bar{C}^*$.

3) C is w*-dentable in Y*.

4) There exists a Banach space Z with a separable dual and an operator $U:Z \to Y$ such that U^* is one-to-one on $\bar{C}^*$ and $U^*(C)$ is norm closed in Z^*.

Proof: 1) => 2) Corollary IV.4 gives that C is a strong norm H_δ-set in $\bar{C}^*$. Proposition III.8 gives then that C is a strong w*-H_δ in $\bar{C}^*$.
2) => 3) follows from Theorem I.8 and 3) => 1) is well known ([4] Theorem 2.3.6). To prove 2) => 4), use first Lemma 0 to get a separable Banach space Y_1 and an embedding $i:Y_1 \longrightarrow Y$ such that $i^*:Y^* \longrightarrow Y_1^*$ is a w*-homeomorphism and isometry on $\bar{C}^*$. Write now $i^*(C) = \bigcap_{n=1}^{\infty} \{f_n \leq \alpha_n\}$ where $f_n \in Y_1^{**}$ and $\alpha_n \in \mathbb{R}$ for each n.

Again, by the Odell-Rosenthal theorem, we can find for each n, a sequence $(g_{n,m})$ in Y_1 such that $g_{n,m}(y^*) \longrightarrow f_n(y^*)$ for each y^* in Y_1^*. We may also suppose without loss of generality that $\|g_{n,m}\| \leq 1$ for all n,m. Define now for each $n \geq 1$ the operator $T_n:\ell_1 \longrightarrow Y_1$ by $T_n(\alpha_m)_m = \sum_m \alpha_m g_{n,m}$ and let $T_0:\ell_2 \longrightarrow Y_1$ be a dense range operator. Consider now the operator $T:\ell_2 \oplus (\sum_n \oplus \ell_1)_{\ell_2} \longrightarrow Y_1$ defined by the sequence of operators $\{T_0, (2^{-n}T_n)_n\}$. Let $T^*:Y_1^* \longrightarrow \ell_2 \oplus (\sum_n \oplus \ell_\infty)_{\ell_2}$ be the adjoint operator which is one-to-one since T_0^* is. Note first that T^* is valued in $\ell_2 \oplus (\sum_n \oplus c)_{\ell_2}$ where c is the subspace of ℓ_∞ consisting of all convergent sequences. Moreover the set $T^* \circ i^*(C)$ is norm closed, since if $y_\ell^* \in i^*(C)$ and $T^* y_\ell^* \longrightarrow z$ then $z = T^* y^*$ for some y^* in $i^*(\bar{C}^*)$.
On the other hand, for each n, $g_{n,m}(y_\ell^*) \longrightarrow g_{n,m}(y^*)$ uniformly in m, hence $f_n(y_\ell^*) \longrightarrow f_n(y^*)$ from which follows that $y^* \in \bigcap_n \{f_n \leq \alpha_n\} = i^*(C)$.

Now T is an operator such that $T^*(Y_1^*)$ is norm separable, hence Stegall's factorization theorem [27] applies and we get a Banach space Z with a separable dual such that $T = S \circ R$ where $R:\ell_2 \oplus (\sum_n \oplus \ell_1)_{\ell_2} \longrightarrow Z$ and $S:Z \longrightarrow Y_1$.

Note finally that S^* is one-to-one and that $\overset{*}{S}\circ i^*(C)$ is norm closed in Z^* since $T^*\circ i^*(C)$ is closed. The operator $U = i\circ S:Z \longrightarrow Y$ clearly verifies claim (4) of the theorem.

4) => follows immediately from the results of Bourgain-Rosenthal (see [4] Theorem 4.1.13).

By combining Theorem II.1 and Theorem IV.5 we get the following strengthening of a result of J. Bourgain [1].

<u>Corollary IV.6</u>: Let Y be a Banach space not containing an isomorphic copy of ℓ_1. Let C be a convex closed bounded subset of Y^* such that $\bar{C}^*$ is w^*-metrizable. If C contains no δ-tree for any $\delta > 0$, then it is contained in the w^*-closed convex hull of its strongly w^*-exposed points.

<u>Proof</u>: By a result of J. Bourgain [1] such a set is necessarily w^*-dentable and separable. It is then a strong w^*-H_δ by Theorem IV.5. The conclusion follows from Theorem II.1.

<u>Theorem IV.7</u>: Let X be a separable Banach space. The following properties are then equivalent:

1) X has the P.C.P.
2) There exists a separable Banach subspace Y of X^* such that X is isometric to a subspace of Y^* with B_X being a strong w^*-G_δ in B_{Y^*}.
3) There exists a separable Banach subspace Y of X^* and a family of norm one vectors $\{y_{n,i}; 1 \le i \le m_n, n \in \mathbf{N}\}$ in Y such that

$$X = \{y^* \in Y^*;\ \lim_n \max_{1\le i\le m_n} |y^*(y_{n,i})| = 0\}$$

<u>Proof</u>: 1) => 2) Apply Proposition IV.2 to the ball of X to get a separable subspace Y of X^* such that if $S:Y \longrightarrow X^*$ is the canonical embedding then

the restriction T of S^* to X is an isometric embedding of X into Y^* and $T(B_X)$ is a w^*-G_δ, w^*-dense in B_{Y^*}. Theorem III.10 then gives that $T(B_X)$ is a strong w^*-G_δ in B_{Y^*}.

2) => 3) This is an adaptation of a proof of Edgar-Wheeler [11]. Let (x_j) be dense in X and let $Y^*\setminus X = \bigcup_{n=1} K_n$ where each K_n is w^*-compact and $d(K_n,X) > \varepsilon_n > 0$. If $y^* \in K_n$, then $d(y^*,X) > \varepsilon_n$ hence there exists g_{y^*} in $X^\perp \subseteq Y^{**}$ with $\|g_{y^*}\| = 1$ and $\langle y^*,g_{y^*}\rangle > \varepsilon_n$. We can find for each $k \in \mathbf{N}$, functionals $g_{y^*,k}$ in B_Y with $\langle x_j, g_{y^*,k}\rangle < \frac{1}{k}$ for $1 \le j \le k$ and $\langle y^*, g_{y^*,k}\rangle > \varepsilon_n$. This shows that, for each k

$$K_n \subseteq \bigcup_{y^* \in K_n} \{y^*;\ \langle y^*,g_{y^*,k}\rangle > \varepsilon_n\}$$

So by the compactness of K_n, there is a finite subcover, say

$$K_n \subseteq \bigcup_i \{y^*; \langle y^*,y_{nik}\rangle > \varepsilon_n\}$$

where i runs over a finite set $\{1,2,\dots,m_{nk}\}$ and $y_{nik} \in B_Y$, $\langle x_j, y_{nik}\rangle < \frac{1}{k}$ for $1 \le j \le k$ and if $y^* \in K_n$ then $\max_{1\le i \le m_{nk}} \langle y^*, y_{nik}\rangle > \varepsilon_n$ for all n.

Define now $p_k(y^*) = \max\{|\langle y^*,y_{nik}\rangle|; 1 \le n \le k,\ 1 \le i \le m_{nk}\}$ and note that if $y^* \notin X$ then $y^* \in K_n$ for some n, so that $p_k(y^*) > \varepsilon_n$ for all $k \ge n$.

On the other hand, if $x \in X$ and $\|x\| = 1$ then given $\varepsilon > 0$ there exists j with $\|x - x_j\| < \varepsilon/2$. If $k > \max\{j,\frac{2}{\varepsilon}\}$, then for $n \le k$ and $1 \le i \le m_{nk}$, we have

$$|\langle x,y_{nik}\rangle| \le \|x - x_j\|\|y_{nik}\| + |\langle x_j,y_{nik}\rangle| \le \frac{\varepsilon}{2} + \frac{1}{k} < \varepsilon.$$

thus $p_k(x) \le \varepsilon$.

3) => 2) It is enough to write $B_{Y^*} \setminus B_X = \bigcup_k L_k$ where each $L_k = \{y^* \in B_{Y^*};\ \max_{1\le i\le m_n} |y^*(y_{n,i})| \ge \frac{1}{k}$ for all $n \ge k\}$. Note that each L_k is w^*-compact and $d(L_k,X) \ge \frac{1}{k}$.

2) => 1) Note that Lemma I.1 gives that every norm closed subset of B_X is a w^*-G_δ, hence by the Baire Category theorem, we get that every norm closed

subset of B_X has points of weak* to norm continuity which are clearly points of weak to norm continuity in X.

Recall that a Banach space X is said to have the <u>Asymptotic norming property</u> (A.N.P.[18]) if there exists a Banach space Y such that X is a subspace of Y* verifying the following property: if (x_n) is a sequence in X and y* is an element of Y* such that: $w^*\text{-}\lim_n(x_n) = y^*$ and $\lim_n \|x_n\| = \|y^*\|$ then $\lim_n \|x_n - y^*\| = 0$.

<u>Theorem IV.8</u>: For a separable Banach space X the following properties are equivalent:

1) X has the R.N.P.

2) There exists a separable subspace Y of X* such that X is isometric to a subspace of Y* with B_X being a strong $w^*\text{-}H_\delta$ in B_{Y^*}.

3) X has the A.N.P.

4) There exists a separable subspace Y of X* such that X is isometric to a subspace of Y* and a family of norm one vectors $\{y_{nm}; n \in \mathbb{N}, m \in \mathbb{N}\}$ in Y such that $X = \{y^* \in Y^*; \underline{\lim}_m y^*(y_{n,m}) \leq 0, \forall n \geq 0\} = \{y^* \in Y^*; \lim_m y^*(y_{n,m}) = 0, \forall n \geq 0\}$

<u>Proof</u>: 1) => 2) Apply Proposition IV.2 to the ball of X to get a separable subspace Y of X* such that if $S: Y \longrightarrow X^*$ is the canonical embedding then $T = S^*|_X$ is an isometric embedding of X into Y* and $T(B_X)$ is a $w^*\text{-}H_\delta$, w*-dense in B_{Y^*}. Theorem III.10 then gives that $T(B_X)$ is a strong $w^*\text{-}H_\delta$ in B_{Y^*}.

2) => 3) Write $Y^* \setminus X = \bigcup_n K_n$ where each K_n is w*-compact convex with $d(K_n, X) \geq \varepsilon_n > 0$. Let (E_n) be an increasing sequence of finite dimensional subspaces of X such that $X = \overline{\bigcup_n E_n}$ and let $|||x||| = \sum_n 2^{-n} d(x, E_n)$. Let now $|\ |_n$ be the semi-norm defined by $|x|_n = d(x, \mathbb{R}_+ K_n) + d(x, -\mathbb{R}_+ K_n)$ and set

$!x! = \sum_n 2^{-n} !x!_n$. Finally, let $\|x\|_1 = \|x\| + |||x||| + !x!$. Note that $\|x\| \le \|x\|_1 \le 7\|x\|$ for each x in Y^* and that $\| \ \|_1$ is w^*-lower semi-continuous hence it is a dual norm on Y^*.

Suppose now that (x_n) is in X and y^* in Y^* such that $\|x_n\|_1 \xrightarrow{n} \|y^*\|$ and $w^*\text{-}\lim_n(x_n) = y^*$. Since each piece of the norm is w^*-lower semi-continuous we get that:

(i) $\|x_n\| \longrightarrow \|y^*\|$

(ii) $|||x_n||| \longrightarrow |||y|||$

(iii) $d(x_n, \mathbf{R}_+K_m) \longrightarrow d(y^*, \mathbf{R}_+K_m)$ for each m.

We claim that $y^* \in X$. Indeed, if not, there exists m such that $y^* \in K_m$ and $\lim_n d(x_n, \mathbf{R}_+K_m) = d(y^*, \mathbf{R}_+K_m) = 0$. We can then suppose that $\|x_n - \lambda_n k_n\| \le \frac{1}{n}$ for some λ_n in $\mathbf{R}_+$ and k_n in K_m. This gives

$$\frac{1}{n} \ge \|x_n - \lambda_n k_n\| = \lambda_n \| \frac{x_n}{\lambda_n} - k_n\| \ge \lambda_n \varepsilon_m.$$

It follows that $\lambda_n \longrightarrow 0$ and $\|x_n\| \longrightarrow 0$; a contradiction.

As noted by Davis-Johnson [8], the fact that $y^* \in X$ coupled with (ii) gives that $\lim_n \|x_n - y^*\| = 0$.

3) => 1) was proved by James-Ho [18]. We sketch an easier proof based on martingales and already used by Davis et Al [7]. Let D be a countable norming set in Y. Let (ϕ_n) be an X-valued bounded martingale. Let ϕ_∞ be a w^*-limit of (ϕ_n) which is valued in Y^*. For each y in D, the real-valued martingale $y(\phi_n)$ converges to $y(\phi_\infty)$ outside a set Ω_y of measure zero. By lemma V.2.9 of [22] the submartingale $\|\phi_n\| = \sup_{y\in D}|y(\phi_n)|$ converges to $\sup_{y\in D}|y(\phi_\infty)| = \|\phi_\infty\|$ outside a set Ω_0 of measure zero. Since X has the A.N.P. with respect to Y, we get that $\lim_n \|\phi_n - \phi_\infty\| = 0$ outside the set $\Omega_0 \cup \bigcup_{y\in D} \Omega_y$ which is of measure zero.

2) => 4) Write $Y^*\setminus X = \bigcup_n K_n$ where each K_n is w*-compact convex and $d(K_n,X)>\varepsilon_n$.

By Theorem I.8, every closed convex bounded subset F of X is w*-dentable, hence for any such set F and any $\varepsilon > 0$, we can find as in Proposition I.2 a sequence (F_j) of closed convex subsets of F and a sequence of w*-open half-spaces (H_j) such that $F \subseteq \bigcup_j (\overline{F_j}^* \cap H_j)$ and $\operatorname{diam}(\overline{F_j}^* \cap H_j) < \varepsilon$. By applying this to any multiple of the ball of X with an integer, we get for any ε_n, a sequence of norm closed convex bounded subsets $(F_k^n)_k$ of X and a sequence of w*-open half spaces $(H_k^n)_k$ such that $X \subseteq \bigcup_k (\overline{F_k^n}^* \cap H_k^n)$ and $\operatorname{diam}(\overline{F_k^n}^* \cap H_k^n) < \frac{\varepsilon_n}{2}$.

Let now f_n be a norm one functional in Y^{**} such that $f_n = 0$ on X and $f_n > \varepsilon_n$ on K_n.

Since the oscillation of f_n on each $\overline{F_k^n}^* \cap H_k^n$ is less than $\frac{\varepsilon_n}{2}$ and $F_k^n \cap H_k^n$ is non-empty, contained in X we get that $|f_n| < \frac{\varepsilon_n}{2}$ on each $\overline{F_k^n}^* \cap H_k^n$. Note now that each H^n is a countable union of w*-closed half-spaces $(H_{k\ell}^n)_\ell$. Therefore, if we write $L_m^n = \operatorname{conv}(\overline{F_k^n}^* \cap H_{k\ell}^n;\ k \text{ and } \ell \leq m)$ we get

(i) $(L_m^n)_m$ is an increasing sequence of w*-compact convex sets in Y^* such that $X \subseteq \bigcup_m L_m^n$ for each n.

(ii) $|f_n| < \frac{\varepsilon_n}{2}$ on each L_m^n since f_n is affine.

It follows that $(L_m^n + \frac{\varepsilon_n}{2} B_{Y^*}) \cap K_n = \emptyset$, hence there exists a functional y_{nm} in Y such that

$$\sup y_{nm}(L_m^n + \frac{\varepsilon_n}{2} B_{Y^*}) < \inf y_{nm}(K_n) = 1.$$

It follows that $\|y_{nm}\| \leq \frac{2}{\varepsilon_n}$ for m large enough, hence the sequence $(y_{nm})_m$

is bounded. We see now that $X \subset \{y^* \in Y^*;\ \lim_m y^*(y_{n,m}) = 0$ for each $n\}$. Indeed, if $x \in X$ and $n \in \mathbf{N}$, then for any $\rho > 0$ there exists M such that $\pm \rho x \in L_m^n$ for all $m \geq M$. Hence $|y_{nm}(\rho x)| < 1$ and $|y_{mn}(x)| < \frac{1}{\rho}$. On the other hand

$$\{y^* \in Y^*;\ \underline{\lim}_m\, y^*(y_{n,m}) \leq 0,\quad \forall\, n \geq 0\} \subset X$$

since $y^*(y_{n,m}) \geq 1$ for every $m \geq 0$ when $y^* \in K_n$, so that finally

$X = \{y^* \in Y^*;\ \underline{\lim}_m\, y^*(y_{n,m}) \leq 0,\ \forall n \geq 0\} = \{y^* \in Y^*;\ \lim_m y^*(y_{n,m}) = 0,\ \forall n > 0\}$.

It simply remains to normalize the vectors $(y_{n,m})$ to finish the proof of 2) => 4).

4) => 2). It is clear that the set $\{y^* \in Y^*;\ \underline{\lim}_m\, y^*(y_m) \leq 0\}$ is a strong w^*-H_δ subset of Y^* whenever $(y_m)_m$ is a bounded sequence in Y. It follows that X is a countable intersection of strong w^*-H_δ sets hence it is a strong w^*-H_δ set in Y^*.

<u>Theorem IV.9</u>: Let Y be a separable Banach space not containing an isomorphic copy of ℓ_1 and let X be a separable subspace of Y^*; the following properties are then equivalent:

1) X has R.N.P.

2) There exists a family of norm one vectors $\{y_{nm};\ n \in \mathbf{N},\ m \in \mathbf{N}\}$ in Y such that $X = \{y^* \in Y^*;\ \lim_m y^*(y_{nm}) = 0$ for all n in $\mathbf{N}\}$ and each sequence $(y_{nm})_m$ is weak Cauchy in Y.

3) There exists a Banach space Z with separable dual and an operator $T: Z \longrightarrow Y$ such that T^* is one-to-one and $T^*(B_X)$ is norm closed in Z^*.

<u>Proof</u>: 1) => 2 By Theorem IV.5, B_X is a strong w^*-H_δ in B_{Y^*}. Corollary III.7 gives that the orthogonal of X in Y^{**} is w^*-separable. That is, there

exists a sequence (y_n^{**}) of norm one vectors in Y^{**} such that $X = \{y^* \in Y^*; y_n^{**}(y^*) = 0 \text{ for all } n \in \mathbb{N}\}$. Since ℓ_1 does not embed in Y, a theorem of Odell-Rosenthal [23] gives for each n, a sequence of functionals $(y_{n,m})_m$ in Y that converges pointwise on Y^* to y_n^{**}. It follows that $X = \{y^* \in Y^*; \lim_m y^*(y_{nm}) = 0 \text{ for all } n \text{ in } \mathbb{N}\}$.

As in Theorem IV.8, 2) implies that B_X is a strong w^*-H_δ in B_{Y^*}, hence X has the Radon-Nikodym property by Theorem I.8.

3) <=> 1) follows from Theorem IV.5 applied to $C = B_X$.

The rest of this section is devoted to the proof of the following

<u>Theorem IV.10</u>: Let X be a Banach space with a countable norming set and let C be a separable bounded closed convex subset of X. If C has R.N.P., then there exists a separable subspace Y of X^* and an isometric embedding $T:X \longrightarrow Y^*$ such that T(C) is a strong w^*-H_δ in Y^*.

For each w^*-compact and convex subset K in X^{**}, define the function

$$\phi(x,K) = \sup\{\inf_{k \in K} \sup_{y \in B_Y} y(x - k);\ \text{over all separable subspaces } Y \text{ of } X^* \}$$

In other words, $\phi(x,K) = \sup\{d_{Y^*}(T^*x, T^*(K));\ Y \text{ is a separable subspace of } X^* \text{ and } T:Y \longrightarrow X^* \text{ is the canonical embedding}\}$. Note that a reasoning similar to the one in Lemma 0 gives:

<u>Lemma IV.11</u>: If C is a separable bounded subset of X and K is a w^*-compact subset of X^{**}, there exists a separable subspace Y of X^* such that $\phi(x,K) = d_{Y^*}(T^*x^*, T^*(K))$ for each x in C, where T is the canonical embedding of Y into X^*.

<u>Lemma IV.12</u>: Let C be a separable bounded closed convex subset of a Banach space X and let H be a w^*-compact convex subset of X^{**} such that $C \cap H = \emptyset$. There exists then a countable ordinal $\gamma(H)$ and a family $(H_\alpha, x_\alpha^*, x_\alpha, r_\alpha)_\alpha$ such

that $(H_\alpha)_\alpha$ is a decreasing family of w*-compact convex subsets of H, (x^*_α) is a family of functionals in X** with $\|x^*_\alpha\| \leq \frac{1}{4}$, $(x_\alpha)_\alpha$ is a family in C and $r_\alpha = \phi(x_\alpha, H_\alpha)$ verifying:

a) $\phi(x, H_\alpha) \geq \phi(x_\alpha, H_\alpha) - x^*_\alpha(x - x_\alpha)$ for all x in C.

b) $H_{\alpha+1} = H_\alpha \setminus V_\alpha$ where $V_\alpha = \{k \in H_\alpha;\ x^*_\alpha(k - x_\alpha) < \frac{3}{8} r_\alpha\}$.

c) If α is an ordinal greater than $\gamma(H)$, we have either $H_\alpha = \emptyset$ or $\phi(x, H_\alpha) = \phi(x, H_{\gamma(H)})$ for all x in C.

<u>Proof</u>: We define the family $(H_\alpha, x^*_\alpha, x_\alpha, r_\alpha)_\alpha$ by transfinite induction, in the following manner:

(i) $H_0 = H$

(ii) If $\alpha = \beta + 1$, let $H_\alpha = H_\beta \setminus V_\beta$. If $H_\alpha \neq \emptyset$, the function $\phi(x, H_\alpha)$ is bounded convex and norm continuous on C, hence the Brondsted-Rockafellar theorem applies and we get x^*_α in X*, $\|x^*_\alpha\| \leq 1/4$ and x_α in C such that $\phi(x, H_\alpha) \geq \phi(x_\alpha, H_\alpha) - x^*_\alpha(x - x_\alpha)$ for all x in C.

(iii) If α is a limit ordinal, let $H_\alpha = \bigcap_{\beta<\alpha} H_\beta$ and get $(x^*_\alpha, x_\alpha, r_\alpha)$ in the same fashion.

Note now that if all the sets (H_α) are non-empty then the family of continuous functions $(\phi(x, H_\alpha))_\alpha$ is increasing and bounded on the separable set C. Hence there exists a countable ordinal $\gamma(H)$ such that $\phi(\cdot, H_\alpha) = \phi(\cdot, H_{\gamma(H)})$ on C for all $\alpha \geq \gamma(H)$.

<u>Proof of Theorem IV.10</u>: Let M be a countable norming set for X. Since C has R.N.P. use Lemma IV.1 to find a sequence of w*-compact convex subsets (K_n) of X** and a sequence of w*-open sets (O_n) in X** determined by a countable set D of functionals in X* with O_n^c convex and w*-closed and such that: If Y is any separable subspace of X* containing $M \cup D$ and if $S: Y \longrightarrow X^*$ is the canonical embedding then $S^*: X^{**} \longrightarrow Y^*$ is an isometry on X verifying:

$$S^*(C) = \bigcap_n (S^*(K_n) \cup S^*(O_n))$$

Let now Y_0 be the smallest Q-vector subspace of X^* generated by $M \cup D$. For each n and each z in X^* let $b(z,n) = \inf\{\lambda; K_n \subseteq \{z < \lambda\}\} = \sup z(K_n)$. We shall construct an increasing sequence $(Y_m)_m$ of countable Q-vector subspaces of X^* containing Y_0 and a countable family

$\{(H_\alpha(n,z), x^*_\alpha(n,z), x_\alpha(n,z), r_\alpha(n,z));\ \alpha \leq \gamma(n,z),\ n \in \mathbf{N},\ z \in \bigcup_m Y_m\}$ such that

a) For each z in Y_m and each n, the family $\{(H_\alpha(n,z), x^*_\alpha(n,z), x_\alpha(n,z), r_\alpha(n,z));\ \alpha \leq \gamma(n,z)\}$ is the one associated to the w*-compact convex set $H_0(n,z) = O_n^c \cap \{z \geq b(z,n)\}$ by Lemma IV.12.

b) Y_{m+1} contains $\{x^*_\alpha(n,z);\ n \in \mathbf{N},\ z \in Y_m\}$.

c) For each z in Y_m, each n in $\mathbf{N}$ and each $\alpha \leq \gamma(n,z)$,

$\phi(x, H_\alpha(n,z)) = d_{Y^*_{m+1}}(S^*_{m+1}x, S^*_{m+1}H_\alpha(n,z))$ where S_m is the canonical embedding of Y_m into X^*.

Start with Y_0 and assume the construction done up to m. For each $(n,z) \in \mathbf{N} \times Y_m$, let $(H_\alpha(n,z); x^*_\alpha(n,z), x_\alpha(n,z), r_\alpha(n,z))_{\alpha \leq \gamma(n,z)}$ be the countable family associated to $H_0(n,z) = O_n^c \cap \{z \geq b(z,n)\}$ by Lemma IV.12. By Lemma IV.11, there exists for each $(n,z) \in \mathbf{N} \times Y_m$ and each $\alpha \leq \gamma(n,z)$ a separable subspace $L = L(n,z,\alpha)$ of X^* such that $\phi(x, H_\alpha(n,z)) = d_{L^*}(S^*x, S^*H_\alpha(n,z))$ for all x in C. Let now Y_{m+1} be the countable Q-vector subspace of X^* generated by Y_m, $\{x^*_\alpha(n,z); n \in \mathbf{N},\ z \in Y_m,\ \alpha \leq \gamma(n,z)\}$ and a countable dense set $D(n,z,\alpha)$ in each $L(n,z,\alpha)$.

Let now $Y = \overline{\bigcup_n Y_n}$ in X^*. Let $S: Y \longrightarrow X^*$ be the canonical embedding. It follows from Lemma IV.1 that $T = S^*|_X$ is an isometric embedding of X into Y^* and that $S^*(C) = \bigcap_n (S^*(K_n) \cup S^*(O_n))$. For each n, we have

$S^*(K_n)^c = \bigcup_{z \in Y_\infty} \{z \geq c(z,n)\}$ where $Y_\infty = \bigcup_n Y_n$ and $c(z,n) \in \mathbf{R}$. Hence

$Y^* \setminus S^*(C) = \bigcup_{z\in Y_\infty} \bigcup_n [\{z \geq c(z,n)\} \cap S^*(O_n)^c]$. To prove that $S^*(C)$ is a strong w^*-H_δ in Y^*, we shall prove that each set $\{z \geq c(z,n)\} \cap S^*(O_n)^c$ is the countable union of w^*-compact convex sets that are a strictly positive distance away from $S^*(C)$.

Note first that $S^*(K_n) \subseteq \{z < c(z,n)\}$ in Y^*, hence $K_n \subseteq \{z < c(z,n)\}$ in X^{**} since S^* is the restriction map. It follows that $b(z,n) \leq c(z,n)$ and $\{z \geq c(z,n)\} \subseteq \{z \geq b(z,n)\}$.

Consider now the set $\{z \geq b(z,n)\} \cap O_n^c = H_0(n,z)$ and let $(H_\alpha(n,z);$ $\alpha \leq \gamma(n,z))$ be the family associated to it by the above construction. Since $z \in Y_\infty$, we have for each α that $\phi(x,H_\alpha) = d_{Y^*}(S^*x, S^*H_\alpha)$ for any x in C.

It follows from Lemma IV.12.a) that $d_{Y^*}(S^*x, S^*H_\alpha) \geq d_{Y^*}(S^*x_\alpha, S^*H_\alpha) - Sx_\alpha^*(S^*x - S^*x_\alpha)$ for all x in C. The same computation as in Theorem III.3 gives that:

(i) $d_{Y^*}\big(S^*(C), S^*(\bar{V}_\alpha^*)\big) \geq \frac{1}{2} d_{Y^*}(S^*x_\alpha, S^*H_\alpha) = \frac{1}{2} r_\alpha > 0$ since $S^*(C) \cap S^*(H_\alpha) = \emptyset$,

(ii) $d(S^*x_\alpha, S^*H_{\alpha+1}) \geq \frac{3}{2} r_\alpha = \frac{3}{2} d(S^*x_\alpha, S^*H_\alpha)$ if $H_\alpha \neq \emptyset$.

Note now that if $H_{\gamma(H)} \neq \emptyset$ then (ii) contradicts Lemma IV.12.c. It follows that $S^*H_0 = \bigcup_{\alpha\leq\gamma(H)} S^*(\bar{V}_\alpha^*)$ and $d(S^*(\bar{V}_\alpha^*), S^*(C)) > 0$. Hence for any z in Y_∞ we have $\{z \geq c(n,z)\} \cap S^*(O_n)^c \subseteq \{z \geq b(n,z)\} \cap S^*(O_n)^c = \bigcup_{\alpha\leq\gamma(H)} S^*(\bar{V}_\alpha^*)$, which finishes the proof of Theorem IV.10.

As a corollary we obtain the following refinement of a result due to J. Bourgain (Theorem 3.5.4 [4]).

<u>Corollary IV.13</u>: Let C be a separable closed bounded convex subset of a Banach space X. If C has RNP, there exists a separable subspace Y of X^* with the following property:

for every norm-closed subset F of C and every subspace Z of X^* containing

Y, the set of functions in Z which strongly expose F is a dense G_δ in Z.

Proof: Choose as Y the separable subspace of X* given by Theorem IV.10 and let Z be any subspace of X* containing Y. Note first that the set $\{x^* \in Z;\ x^*$ strongly exposes $F\}$ is always a G_δ in Z. To prove that it is dense, take any $x^* \in Z$, $x^* \notin Y$ and consider $Y_1 = Y \oplus \mathbb{R}x^*$. According to Theorem II.1 (and using also Lemma I.1 to go from C to F) it is enough to show that C is a strong w^*-H_δ in Y_1^*.

Let Q be the projection from Y_1^* onto Y^*, D and D_1 respectively the w^*-closures of C in Y^* and Y_1^*. Note that $Q(D_1) = D$ and that $\operatorname{Ker}Q = L$ is one-dimensional. Theorem IV.10 asserts that C is a strong w^*-H_δ in Y^*, therefore $D \setminus C = \bigcup_n K_n$, with K_n w^*-compact, convex and $d(K_n,C) > 0$; if we set $\tilde{K}_n = D_1 \cap Q^{-1}(K_n)$ we get

$$D_1 \setminus Q^{-1}(C) = \bigcup_n \tilde{K}_n \text{ with } \tilde{K}_n \text{ } w^*\text{-compact, convex}$$

and $d(\tilde{K}_n,C) > 0$. Observe now that $D_1 \cap Q^{-1}(C) \subset C + L$ is norm separable and apply Lemma I.1 to get that C is a strong w^*-H_δ in $D_1 \cap Q^{-1}(C)$, thus finally C is a strong w^*-H_δ in D_1.

V. ON w^*-H_δ SETS IN FUNCTION SPACES:

In this section we study the special relations between w^*-G_δ, w^*-H_δ and w^*-compact subsets of Banach lattices. We start with the case of subsets of spaces of measures on a compact Hausdorff space.

We need the following lemma due to J. Bourgain ([2] Lemma 5.3).

Lemma V.1: Let C be a bounded convex subset of a dual Banach space Y* and let U be a non-empty relatively w^*-open subset of C. Then there exist a finite number $S_1,\ldots,S_p$ of w^*-open slices of C and positive scalars $\lambda_1,\ldots,\lambda_p$ with $\sum_{i=1}^{p}\lambda_i = 1$ so that $\sum_{i=1}^{p}\lambda_i S_i \subseteq U$.

Proof: Denote by E the set of extreme points of $\bar{C}^*$. Let x be any element of U. There is a convex w^*-neighborhood V of 0 such that $(x + 2V) \cap C \subseteq U$. Since $x \in \overline{\text{conv}}^*(E)$, we can find $e_1,\ldots,e_p$ in E and positive scalars $\lambda_1\ldots,\lambda_p$ with $\sum_{i=1}^{p}\lambda_i = 1$ such that $\sum_{i=1}^{p}\lambda_i e_i \in x + V$. By the extremality of each e_i, we can find a w^*-open slice $\tilde{S}_i$ of $\bar{C}^*$ so that $\tilde{S}_i \subseteq e_i + V$. Then $S_i = \tilde{S}_i \cap C$ is a slice of C such that

$$\sum_{i=1}^{p}\lambda_i S_i \subseteq \sum_{i=1}^{p}\lambda_i (e_i + V) \cap C \subseteq (x + 2V) \cap C \subseteq U. \quad \text{Q.E.D.}$$

Let now K be metrizable compact Hausdorff space and let μ be a Radon probability measure on K. For any bounded subset A of $L_1(K,\mu)$, define the modulus of equi-integrability of A to be

$$\delta(A) = \lim_{\varepsilon \to 0} \sup \left\{ \int_E |f| d\mu ; f \in A \text{ and } \mu(E) \leq \varepsilon \right\}$$

The following lemma summarizes the properties of δ. The proof is left to the interested reader.

Lemma V.2: For any bounded subset A of $L_1(K,\mu)$ we have:

(i) $\delta(A + g) = \delta(A)$ for each g in $L_1(K,\mu)$

(ii) $\delta(A) \leq \text{diam}(A)$

(iii) For each λ in the w*-closure $\bar{A}^*$ of A in $\mathcal{M}(K)$, we have $d(\lambda, L^1) \leq \delta(A)$

(iv) For each family $(A_i)_{i=1}^n$ of bounded subsets of L_1 and each sequence $(\theta_i)_{i=1}^n$, $\theta_i \geq 0$ with $\sum_{i=1}^n \theta_i = 1$ we have

$$\delta\left(\sum_{i=1}^n \theta_i A_i\right) = \sum_{i=1}^n \theta_i \delta(A_i)$$

Remark: The proof of (iii) is an immediate application of the subsequence splitting lemma. [31]

Lemma V.3: Let C be a convex bounded subset of L_1 which is a w*-G_δ subset of $\mathcal{M}(K)$, then for each $\varepsilon > 0$, there exists a w*-open slice S of C such that for each $\lambda \in \bar{S}^*$, $d(\lambda, L_1) \leq \varepsilon$.

Proof: Since C is a norm separable w*-G_δ subset of $\mathcal{M}(K)$, the Baire category theorem insures the existence of a w*-open set U in $\mathcal{M}(K)$ such that $C \cap U \neq \phi$ and $\text{diam}(C \cap U) \leq \varepsilon$. By Lemma V.1, there exists w*-open slices $S_1, \ldots, S_n$ of C such that

$$\sum_{i=1}^n \theta_i S_i \subseteq C \cap U \text{ where } \theta_i \geq 0 \ \forall i \text{ and } \sum_{i=1}^n \theta_i = 1.$$

Note now that

$$\sum_{i=1}^n \theta_i \delta(S_i) \leq \delta\left(\sum_{i=1}^n \theta_i S_i\right) \leq \delta(C \cap U) \leq \text{diam}(C \cap U) \leq \varepsilon.$$

Hence there exists S_{i_o} such that $\delta(S_{i_o}) \leq \varepsilon$. It follows that for each $\lambda \in \bar{S}^*_{i_o}$, $d(\lambda, L_1) \leq \varepsilon$.

Lemma V.4: Let C be a convex bounded subset of $L_1(K,\mu)$ which is also a w*-G_δ in $\mathcal{M}(K)$. Then, there exists a sequence (K_n) of w*-compact convex subsets of $\mathcal{M}(K)$ and a sequence of w*-open subsets (O_n) consisting of countable union of w*-open half-spaces of $\mathcal{M}(K)$ such that:

$$C \subseteq \bigcap_n (K_n \cup O_n) \subseteq L_1(K,\mu) \cap \bar{C}^*.$$

Proof: Fix $\varepsilon > 0$ and define inductively a decreasing family of relatively w*-closed subsets of C, which are w*-G_δ in $\mathcal{M}(K)$, in the following way:

(i) $F_0 = C$

(ii) If $\alpha = \beta + 1$ and $F_\beta \neq \phi$ use Lemma V.3 to find a w*-open slice S_β of F_β such that for each $\lambda \in \bar{S}^*_\beta$, $d(\lambda, L_1) \leq \varepsilon$. Set $F_\alpha = F_\beta \setminus S_\beta$

(iii) IF α is a limit ordinal, let $F_\alpha = \bigcap_{\beta<\alpha} F_\beta$

Since the ball of $\mathcal{M}(K)$ is w*-metrizable, there exists $\gamma_\varepsilon < \Omega$ (the first uncountable ordinal) such that $F_{\gamma_\varepsilon} = \phi$. Let K_α be the w*-closure of F_α in $\mathcal{M}(K)$ and let H_α be the w*-open half-space such that $S_\alpha = H_\alpha \cap F_\alpha$. It is clear that

$$C \subseteq \bigcap_{\alpha \leq \gamma_\varepsilon} (K_\alpha \cup \bigcup_{\beta<\alpha} H_\beta)$$

Moreover, if x belongs to the right-hand side, then $x \in K_\beta \cap H_\beta$ for some $\beta \leq \gamma_\varepsilon$, hence $x \in \bar{S}^*_\beta$ and $d(x, L_1) \leq \varepsilon$. It follows that if we repeat the construction for each $\varepsilon = \frac{1}{n}$ we would get:

$$C \subseteq \bigcap_n \bigcap_{\alpha \leq \gamma_n} (K_{\alpha,n} \cup \bigcup_{\beta<\alpha} H_{\beta,n}) \subseteq \bar{C}^* \cap L_1(K,\mu).$$

Now, we can prove the following

Theorem V.5: Let K be a metrizable compact Hausdorff space and let C be a separable closed convex subset of $\mathcal{M}(K)$. Then C is a w*-H_δ set if and only if it is a w*-G_δ set.

Proof: Since C is norm separable, we can assume that $C \subseteq L_1(K,\mu)$ for some Radon probability measure μ on K. If C is a w*-G_δ, apply Lemma V.5 to get a sequence (K_n) of w*-compact convex subsets of $\mathcal{M}(K)$ and a sequence of w*-open sets (O_n) consisting of countable union of w*-open half-spaces of $\mathcal{M}(K)$ such that

$$C \subseteq D = \bigcap_n (K_n \cup O_n) \subseteq L_1(K,\mu) \cap \bar{C}^*.$$

Write now each K_n as a countable intersection of w*-open half-spaces in $\mathcal{M}(K)$ to conclude that D is a w*-H_δ set. Since D is norm separable and C is a norm closed subset of D, Lemma I.1.b gives that C is a w*-H_δ set in $\mathcal{M}(K)$.

Remark V.6: The assumption of norm separability on C is crucial in Theorem V.5, since the set $\mathcal{P}_c[0,1]$ of continuous measures on [0,1] is a w*-G_δ in $\mathcal{P}[0,1]$ but not a w*-H_δ set since it contains no extreme points.

Now we turn to more general Banach lattices not containing an isomorphic copy of c_0. We shall prove the following:

Theorem V.7: Let X be a Banach lattice not containing an isomorphic copy of c_0. Let C be a closed convex subset of X. Then C has the P.C.P if and only if it has the R.N.P.

For the proof of the above theorem (and the next one) we need the following proposition.

Proposition V.8: Let X be a separable Banach lattice not containing an isomorphic copy of c_0 and let C be a closed bounded convex subset of X. If C has the P.C.P, then there exists a separable Banach lattice Y, a metrizable compact Hausdorff space Ω, a Radon probability measure μ on Ω and a dense range lattice homomorphism $S: C(\Omega) \longrightarrow Y$ such that:

(i) X is lattice isometric to a closed order ideal of Y^* in such a way that C is a w^*-G_δ in Y^*.

(ii) The restriction of the operator $S^*: Y^* \longrightarrow \mathcal{M}(\Omega)$ to X maps it into $L_1(\Omega,\mu)$ in such a way that:

(a) $S^{*-1}(L_1(\Omega,\mu) \subseteq X$

(b) $L_\infty(\Omega,\mu)$ is an order ideal of $S^*(X)$.

Proof: By Lemma IV.1, C can be written as $\bigcap_n (K_n \cup O_n)$ where each K_n is w^*-compact in X^{**} and each O_n is a countable union of w^*-open elementary neighborhoods of X^{**}. Let Y be the smallest separable sublattice of X^* containing a norming subset for X and all the functionals determining the w^*-open sets (O_n). let T be the canonical injection of Y into X^*. We get as in Proposition IV.2 that the restriction of the quotient map $T^*: X^{**} \to Y^*$ to the space X is an isometry that maps the set C into a w^*-G_δ of Y^*. On the other hand, T is a lattice isometry, hence T^* is interval preserving [21]; since X is an order ideal in X^{**} ([20] Theorem I.b.16), we get that $T^*(X)$ is an order ideal in Y^*. We shall now identify X with $T^*(X)$ which clearly verifies the assertion (i) of the proposition.

Let now u_o be a norm one quasi-interior point of Y and let $C(\Omega_o)$ be the A-M space obtained by norming the vector lattice $\bigcup_n [-nu_o, nu_o]$ with the gauge of the order interval $[-u_o, u_o]$. Note that the canonical injection $S: C(\Omega_o) \to Y$ is a dense range, lattice homomorphism which is also interval preserving. Let $C(\Omega)$ be the separable sublattice of $C(\Omega_o)$ generated by a countable Y-dense subset of $[-u_o, u_o]$ in view of the separability of Y.

Note now that $S^*: Y^* \to \mathcal{M}(\Omega)$ is an interval preserving lattice homomorphism which is also one-to-one. Since X is separable, consider a quasi-interior point x_o of X such that $u_o(x_o) = 1$ and set $\mu = S^*x_o$. It is clear that $L_\infty(\Omega,\mu) \subseteq S^*(X) \subseteq L_1(\Omega,\mu)$.

To prove (ii)(a) note that if $y^* \in Y^*_+$ is such that $S^*y^* \in L_1(\Omega,\mu)$ then $S^*y^* \wedge n = S^*(y^* \wedge nx_o)$ converges in L_1 to S^*y^*, hence $S^*(y^* \wedge nx_o) \xrightarrow{w^*} S^*y^*$

and $y^* \wedge nx_o$ w*-converges to y^* since S^* is a w*-homeomorphism on the bounded subsets of Y^*. But the sequence $(y^* \wedge nx_o)$ is increasing and norm-bounded in X hence it converges in norm to an element in X which must be y^*, since c_0 does not embed in X ([20] Theorem I.c.4).

Proof of Theorem V.7: Since both properties (P.C.P) and (R.N.P) are separably determined ([10],[11]), we can assume that C (and hence X) is norm separable. Apply Proposition V.8 to find Y such that C is a w*-G_δ in Y^*. Note that $C_1 = S^*(C)$ is a w*-G_δ in $\mathcal{M}(\Omega)$ which is contained in $L_1(\Omega,\mu)$. Apply Lemma V.4 to C_1 to get that

$$C_1 \subseteq \bigcap_n (K_n \cup O_n) \subset \bar{C}_1^* \cap L_1(\Omega,\mu)$$

where each K_n is w*-compact and convex and each O_n is a countable union of w*-open half-spaces in $\mathcal{M}(\Omega)$. Since S^* is one-to-one and $S^{*-1}(L_1(\Omega,\mu)) \subseteq X$, we get that

$$C \subseteq D = \bigcap_n (S^{*-1}(K_n) \cup S^{*-1}(O_n)) \subseteq X.$$

But $S^{*-1}(K_n)$ is w*-compact and convex in Y^* and $S^{*-1}(O_n)$ is a countable union of w*-open half-spaces in Y^* hence D is a w*-H_δ set in Y^*. Since X is separable and C is norm closed, Lemma I.1 gives that C is a w*-H_δ set in Y^*, hence it is an R.N.P. set.

Now, we can prove the following extension of a theorem of Talagrand [29].

Theorem V.9: Let X be a separable Banach lattice. The following properties are then equivalent:

1) X has the P.C.P.

2) X has the R.N.P.

3) X is isometric to the dual of a Banach lattice.

Proof: Note that all these properties imply that c_0 does not embed in X. Hence, in view of Theorem V.7 it is enough to prove 2) => 3). Apply proposition V.8 to $C = B_X$ to get Y such that X is isometric to an order ideal of Y* in such a way that $Y^* \setminus X = \bigcup_n K_n$ where each K_n is w*-compact and convex. Let $S:C(\Omega) \longrightarrow Y$ be as in the conclusion of the proposition: that is $L_\infty(\Omega,\mu) \subseteq S^*(X) \subseteq L_1(\Omega,\mu)$.

We first show that for all $\varepsilon > 0$, there exists a measurable subset $\Omega_\varepsilon \subseteq \Omega$ with $\mu(\Omega_\varepsilon) \geq 1 - \varepsilon$ such that the set $B_\varepsilon = \{x \in B_X;\ S^*x = 0 \text{ on } \Omega \setminus \Omega_\varepsilon\}$ is w*-compact in B_{Y^*}. Indeed, for each n, we have $(\frac{2^n}{\varepsilon})B_{L_\infty} \subseteq S^*(X)$ hence it is w*-compact convex and is disjoint from $S^*(K_n)$. There exists then $\phi_n \in C(\Omega)$ with $\phi_n \leq 1$ on $(\frac{2^n}{\varepsilon})B_{L_\infty}$ and $\phi_n > 1$ on K_n.

It follows that $\|\phi_n\|_1 \leq \varepsilon\, 2^{-n}$ and if we set $\Omega_n = \{\omega \in \Omega; |\phi_n(\omega)| \leq 1\}$ we get $\mu(\Omega \setminus \Omega_n) \leq \varepsilon\, 2^{-n}$.

Let now (x_k) in B_X such that $S^*x_k = 0$ on $\Omega \setminus \Omega_n$ and suppose that $\text{w*-}\lim_k (x_k) = y^* \in Y^*$. We have:

$$| < S^*x_k, \phi_n > | = | \int_{\Omega_n} (S^*x_k)\phi_n d\mu | \leq \|S^*x_k\|_1 \leq \|x_k\| \leq 1.$$

Hence $< S^*y^*, \phi_n > \leq 1$ and $y^* \notin K_n$.

If now $\Omega_\varepsilon = \bigcap_{n \geq 1} \Omega_n$, we get that $\mu(\Omega_\varepsilon) \geq 1 - \varepsilon$ and if (x_k) is a sequence in B_X such that $S^*x_k = 0$ on $\Omega \setminus \Omega_\varepsilon$ and $\text{w*-}\lim x_k = y^*$ in Y* then $y^* \notin \bigcup_n K_n$ hence $y^* \in B_X$ and the claim is proved.

If now P_ε is the band projection on the subideal of X consisting of the functions supported on Ω_ε, we get that $P_\varepsilon(X)$ is a dual space and $(I - P_\varepsilon)X$ is also an R.N.P. Banach lattice, hence the same reasoning applies to it. We then get by a standard exhaustion argument a sequence of disjoint band projections (P_n) such that each $P_n(X)$ is a dual space and since c_0 does not embed in X, the latter is a boundedly complete disjoint sum of $(P_n(X))_n$ hence it is a dual Banach lattice.

VI. OPTIMIZATION ON STRONG w^*-G_δ SETS:

In this section we shall describe a method for associating w^*-H_δ sets to w^*-G_δ sets. This procedure will allow us to study some optimization problems on G_δ-sets by reducing them to the questions that we dealt with in the preceding sections.

Let Y be a Banach space and let D be a w^*-compact w^*-metrizable subset of Y^*. Denote by Lip(D) the space of the restrictions to D of norm-Lipschitz functions on Y^* that are w^*-continuous on D; the norm on Lip(D) being define by $\|f\|_{BL} = \max\|f\|_L, \|f\|_\infty)$ where

$$\|f\|_L = \sup\left\{ \frac{|f(x)-f(y)|}{\|x-y\|} \; ; \; x \neq y \quad x,y \in Y^* \right\} \quad \text{and} \quad \|f\|_\infty = \sup\{|f(x)| \; ; \; x \in D\}.$$

Note that $Y \subseteq \mathrm{Lip}(D) \subseteq C(D)$ and by the Stone Weierstrass theorem Lip(D) is norm dense in C(D). If now $\mathcal{P}(D)$ denotes the set of probabilities in the dual of C(D), then the adjoint map from $C(D)^*$ into $\mathrm{Lip}(D)^*$ is an homeomorphism for the respective w^*-topologies on the set $\mathcal{P}(D)$. Therefore $\mathcal{P}(D)$ is w^*-compact and w^*-metrizable in $\mathrm{Lip}(D)^*$.

If now C is a norm closed and separable subset of D then C is a w^*-$K_{\sigma\delta}$ set ([4] Lemma 4.3.6); in particular the norm-Borel and w^*-Borel σ-fields on C coincide. We shall denote by $\mathcal{P}(C)$ the set of probabilities on D that are supported on C.

The main result of this section is the following:

Theorem VI.1: Let Y be a Banach space and let D be a weak*-compact w^*-metrizable subset of Y^*. Suppose C is a norm closed and separable subset of D, then $\mathcal{P}(C)$ is a norm closed separable subset of $\mathcal{P}(D)$ and the following assertions are equivalent:

(a) C is a w^*-G_δ (resp strong w^*-G_δ) subset of D in Y^*.

(b) $\mathcal{P}(C)$ is a w*-H_δ (resp strong w*-H_δ) subset of $\mathcal{P}(D)$ in Lip(D)*.

In the following lemmas, D will always be a weak*-compact w*-metrizable subset of the dual of a Banach space Y and C will be a separable norm-closed subset of D. We shall also assume that $\operatorname{diam}(D) \leq 1$.

Lemma VI.2: The set $\mathcal{P}(C)$ is a separable norm closed subset of Lip(D)*

Proof: Suppose μ is in the norm closure of $\mathcal{P}(C)$ in Lip(D)*. Then $\mu \in \mathcal{P}(D)$ and one can find a sequence (μ_n) in $\mathcal{P}(C)$ such that $\|\mu-\mu_n\|_{\text{Lip}(D)^*} \leq 4^{-n-1}$. Because C is Polish in the norm topology each μ_n is tight; hence there exists for each n a norm compact set $K_n \subseteq C$ such that $\mu_n(C \setminus K_n) \leq 4^{-n-1}$. Note now that the functions $\phi_n : x \to d(x,K_n)$ can be obtained by iterated pointwise limits of functions in the unit ball of Lip(D). (See Lemma VI.11). Hence by Lebesgue's theorem we have:

$$\left|\int d(x,K_n)(d\mu-d\mu_n)\right| \leq 4^{-n-1}.$$

But $0 \leq \int d(x,K_n)\,d\mu_n \leq 4^{-n-1}$, hence $\int d(x,K_n)\,d\mu \leq 4^{-n}$ and $\mu\{x; d(x,K_n) \geq 2^{-n}\} \leq 2^{-n}$. It follows that $\mu(B_m) \geq 1 - 2^{-m}$ where $B_m = \bigcap_{n>m} \{x; d(x,K_n) \leq 2^{-n}\}$. If now $x \in B_m$, there exists for each $n > m$, $k_n \in K_n$ such that $d(x,k_n) \leq 2^{-n}$; that is $x \in C$ since it is norm closed. Hence $B_m \subseteq C$ and μ is supported by C.

Note finally that the atomic measures supported by a countable dense set in C, are dense in $\mathcal{P}(C)$ for the norm of Lip(D)*. Hence $\mathcal{P}(C)$ is norm separable.

Lemma VI.3: If $\mathcal{P}(C)$ is a w*-H_δ set in $\mathcal{P}(D)$, then for each $\varepsilon > 0$ there exists a w*-neighborhood V in Y* such that $V \cap C \neq \emptyset$ and $\operatorname{diam}(V \cap C) \leq \varepsilon$.

Proof: By Theorem I.8, $\mathcal{P}(C)$ is then w*-dentable in Lip(D)*, hence there exists a w*-denting point μ in $\mathcal{P}(C)$. Note first that μ is a Dirac measure of the form δ_x for some x in C since it is obviously an extreme point.

Moreover, there exists $\phi \in \mathrm{Lip}(D)$ and $\tau > 0$ such that $|\phi(\mu) - \phi(\delta_x)| < \tau \Rightarrow \|\mu - \delta_x\|_{\mathrm{Lip}(D)^*} \leq \varepsilon$ for all $\mu \in P(C)$. Note now that if $y \in C$ then $\delta_y \in P(C)$ and $\|\delta_y - \delta_x\|_{\mathrm{Lip}(D)^*} = \|y-x\|$ hence the relative w*-neighborhood of Y* defined by $V = \{y \in D; |\phi(y) - \phi(x)| < \tau\}$ verifies the claim of the Lemma.

<u>Lemma VI.4</u>: If C_1 is a relatively w*-closed subset of C and if $P(C)$ is a w*-H_δ in $P(D)$ then $P(C_1)$ is also a w*-H_δ in $P(D)$.

<u>Proof</u>: Suppose that $C_1 = C \cap K$ where K is a w*-compact subset of D. We have that $D \setminus K = \bigcup_n K_n$ where (K_n) is an increasing sequence of w*-compact subsets of D. Write now that

$P(C) \setminus P(C_1) \subseteq \bigcup_n \tilde{K}_n$ where $\tilde{K}_n = \{\mu \in P(D); \mu(K_n) \geq \frac{1}{n}\}$ is w*-compact and convex.

This immediately implies that $P(C_1)$ is a w*-H_δ in $P(D)$.

<u>Proof of Theorem VI.1</u>: a) => b) Suppose that C is a w*-G_δ in D; that is $D \setminus C = \bigcup_n K_n$ where (K_n) is an increasing sequence of w*-compacts subsets of D. It is clear that $P(D) \setminus P(C) = \bigcup_n \tilde{K}_n$ where each $\tilde{K}_n = \{\mu \in P(D); \mu(K_n) \geq \frac{1}{n}\}$ is w*-compact and convex.

Suppose now that C is a strong w*-G_δ subset of D. That is $D \setminus C = \bigcup_n K_n$ where (K_n) is an increasing sequence of w*-compact subsets of D such that $d(K_n, C) > 0$. We shall now prove that each $\tilde{K}_n = \{\mu \in P(D); \mu(K_n) \geq \frac{1}{n}\}$ is also far from $P(C)$ for the norm of $\mathrm{Lip}(D)^*$. Indeed, for any $\mu \in P(C)$ and $\nu \in P(D)$ such that $\|\mu-\nu\|_{\mathrm{Lip}(D)^*} \leq 4^{-\ell-1}$ we can obtain as in the proof of Lemma VI.2, a norm compact subset $K \subseteq C$ such that $\nu\{x; d(x,K) \geq 2^{-\ell}\} \leq 2^{-\ell}$. If now $2^{-\ell} < \min(d(K_n, C), \frac{1}{n})$ then $\nu \notin \tilde{K}_n$. Hence $d(\tilde{K}_n, P(C)) > 0$ for each n.

b) => a). Suppose $P(C)$ is a w*-H_δ in $P(D)$. Lemmas VI.4 and VI.3 give then the following: For every relatively w*-closed subset C_1 of C and every $\varepsilon > 0$, there exists a w*-neighborhood V of D such that $V \cap C_1 \neq \phi$ and $\text{diam}(V \cap C_1) \leq \varepsilon$. Use this property to define inductively a decreasing family (C_α) of w*-relatively closed subsets of C in the following way: For a fixed $\varepsilon > 0$ let $C_0 = C$. If $\alpha = \beta + 1$ and $C_\beta \neq \phi$, find a w*-open neighborhood V_β such that $V_\beta \cap C_\beta \neq \phi$ and $\text{diam}(V_\beta \cap C_\beta) \leq \varepsilon$. Set $C_\alpha = C_\beta \setminus V_\beta$. If α is a limit ordinal, let $C_\alpha = \bigcap_{\beta<\alpha} C_\beta$. Since C is norm separable, there exists $\gamma_\varepsilon < \Omega$ (the first uncountable ordinal) such that $C_{\gamma_\varepsilon} = \phi$. Let K_α be the w*-closure of C_α in D. We obtain as in the proof of Proposition I.2 that

$$C \subseteq \bigcap_{\alpha \leq \gamma_\varepsilon} (K_\alpha \cup \bigcup_{\beta<\alpha} V_\beta)$$

and if one repeats the construction for each $\varepsilon = \frac{1}{n}$ then

$$C = \bigcap_n \bigcap_{\alpha<\gamma_n} (K_{\alpha,n} \cup \bigcup_{\beta<\alpha} V_{\beta,n})$$

Since the set D is w*-metrizable, we get that C is a w*-G_δ in D.

Suppose now that $P(C)$ is a strong w*-H_δ subset of $P(D)$ in $\text{Lip}(D)^*$. We get from the above that $D \setminus C = \bigcup_n K_n$ where each K_n is w*-compact. To show that each K_n can be written as a countable union of w*-compact sets $K_{n,m}$ such that $d(K_{n,m}, C) > 0$, note that for each w*-compact set K, the function $\nu \to \int d(x,K)d\nu(x)$ is an affine w*-lower semi-continuous function on $P(C)$, hence we can use Theorem II.11 to find a function $\phi \in \text{Lip}(D)$ with $\|\phi\| \leq \frac{1}{4}$ and such that the function $\nu \to \int (d(x,K) + \phi(x))d\nu(x)$ attains its minimum on $P(C)$ necessarily on a Dirac measure δ_{x_0} for some x_0 in C. In other words, one can find for each w*-compact set $K \subseteq D$, a point x_0 in C and a function $\phi \in \text{Lip}(D)$ with $\|\phi\| \leq \frac{1}{4}$ such that $d(x_0,K) + \phi(x_0) \leq d(x,K) + \phi(x)$ for each $x \in C$.

Let now $V = \{x \in D; \phi(x) < \phi(x_0) + \frac{3}{8} d(x_0,K)\}$ and note that if $x \in C$ and $\phi(x) \leq \phi(x_0) + \frac{1}{2} d(x_0,K)$ then $d(x,K) \geq \frac{1}{2} d(x_0,K)$. On the other hand, if $x \in C$,

$\phi(x) > \phi(x_0) + \frac{1}{2} d(x_0,K)$ and $k \in \bar{V}^* \cap K$, we get from the Lipschitz property of ϕ:

$$\|x-k\| \geq 4|\phi(x) - \phi(k)| \geq 4\big||\phi(x) - \phi(x_0)| - |\phi(k) - \phi(x_0)|\big| \geq \frac{1}{2} d(x_0,K).$$

Similarly, $d(x_0,K \setminus V) \geq \frac{3}{2} d(x_0,K)$. In particular, if $x_0 \notin K$ then $K \cap V \neq \emptyset$.

It follows that for each w*-compact subset $K \subseteq D$ such that $K \cap C = \phi$, there exists a w*-open neighborhood V such that $V \cap K \neq \phi$ and $d(C,\bar{V}^* \cap K) > 0$. Lemma III.1 applies and we get a countable partition of K into w*-compact subsets with strictly positive distances from C.

By applying this process to each K_n, we obtain that C is a strong w*-G_δ subset of D.

We are now ready to study the non-convex analogues of Theorem II.11. First let us say that a bounded subset C of a dual space Y* has the <u>*-Ekeland property</u> if for every bounded below norm-lower semi-continuous function ϕ on C and every $\varepsilon > 0$, there exists a w*-continuous and norm-lipschitz function h with $\|h\|_{Lip(\bar{C}^*)} \leq \varepsilon$ such that $\phi + h$ attains its minimum on C. We shall say that x* is a <u>strong w*-peak</u> point in C iff there exists a w*-continuous and norm lipschitz function ϕ on $\bar{C}^*$ such that:

(a) $\phi(x^*) > \phi(z^*)$ for all z* in C, $z^* \neq x^*$.

(b) If (x_n^*) is a sequence in C such that $\phi(x_n^*) \to \phi(x^*)$ then (x_n^*) norm converges to x*.

In this case ϕ is said to be <u>a strongly exposing function for C at x*</u>.

Note first that every strong w*-peak point of a set C is necessarily a point of weak* to norm continuity relative to C. Moreover if C is a separable w*-G_δ in $\bar{C}^*$ and if the latter is w*-metrizable, then the above discussion shows that x is a point of weak* to norm continuity (resp a strong w*-peak point) for C if and only if δ_x is a w*-denting point (resp a strongly w*-exposed point) for $\mathcal{P}(C)$ when considered as a subset of $Lip(\bar{C}^*)$. The following is now an immediate application of the above discussion and Theorem II.1.

Proposition VI.5: Let Y be a Banach space and let C be a bounded separable subset of Y* such that $\bar{C}^*$ is w*-metrizable. Suppose C is a strong w*-G_δ in $\bar{C}^*$ then the set of strongly exposing functions for C is a dense G_δ in Lip($\bar{C}^*$). Furthermore, the set of strong w*-peak points of C is w*-dense in C.

Proof: It is enough to notice that in this case $P(C)$ is a separable strong w*-H_δ in $P(\bar{C}^*)$ when considered as a subset of the dual of Lip($\bar{C}^*$).

Lemma VI.6: Let Y be a Banach space and let C be a closed bounded separable subset of Y* such that $D = \bar{C}^*$ is w*-metrizable. Let $\{\mu_n, \mu\}_n$ be a sequence of measures in $P(C)$ such that $\lim_n \|\mu_n - \mu\| = 0$ in the dual of Lip(D), then $\lim_n \mu_n(f) = \mu(f)$ for each norm continuous and bounded function f on C.

Proof: By a theorem of Dudley ([4] Proposition 6.3.5), it is enough to show that the sequence $\{\mu_n\}$ is norm-tight: that is for each $\eta > 0$, there exists a norm compact set K in C such that $\mu_n(K) > 1 - \eta$ for all n. For that, assume diam (C) ≤ 1 and use the fact that each measure μ_ℓ is norm-tight, to find for each n a norm compact subset K_ℓ^n such that $\mu_\ell(C \setminus K_\ell^n) \leq \frac{\eta}{2^{n+1}} \cdot \frac{1}{n}$. Let now m be a large enough integer such that $\|\mu_\ell - \mu_m\| \leq \frac{\eta}{2^{n+1}} \cdot \frac{1}{n}$ for all $\ell \geq m$.

Note that the function $\Phi_m(x) = d(x, K_m^n)$ can be obtained by iterated pointwise limits of functions in the unit ball of Lip(D), hence by Lebesgue's theorem we have:

$$\left| \int d(x, K_m^n)(d\mu_\ell - d\mu_m) \right| \leq \frac{\eta}{2^{n+1}} \cdot \frac{1}{n} \text{ for all } \ell \geq m.$$

But $0 \leq \int d(x, K_m^n) d\mu_m \leq \frac{\eta}{2^{n+1}} \cdot \frac{1}{n}$, hence $\int d(x, K_m^n) d\mu_\ell \leq \frac{\eta}{2^n} \cdot \frac{1}{n}$ and

$\mu_\ell\{x; d(x,K_m^n) \geq \frac{1}{n}\} \leq \frac{\eta}{2^n}$. It follows that for each $\ell \geq m$,

$\mu_\ell(K_m^n + \frac{1}{n} B_{Y*}) \geq 1 - \frac{\eta}{2^n}$. Let now $L_n = K_1^n \cup K_2^n \cup \ldots \cup K_m^n$. It is norm compact and

$\mu_\ell(L_n + \frac{1}{n} B_{Y*}) \geq 1 - \frac{\eta}{2^n}$ for all ℓ.

Note now that $K = \bigcap_n (L_n + \frac{1}{n} B_{Y*})$ is norm-compact and that $\mu_\ell(K) \geq 1 - \eta$ for all ℓ. Q.E.D.

Now we can show the following:

<u>Theorem VI.7</u>: Let Y be a Banach space and let C be a bounded separable subset of Y* such that $D = \bar{C}^*$ is w*-metrizable. Suppose C is a w*-G_δ in D then the following properties are equivalent:

1) C is a strong w*-G_δ in D.

2) C has the *-Ekeland property.

<u>Proof</u>: To prove that 1) => 2), note first that Theorem VI.1 gives that $P(C)$ is a separable strong w*-H_δ in $P(D)$ when considered as a subset of Lip(D)*. Let now ϕ be a bounded norm-lower semi-continuous function on C. It is then the supremum of a sequence (ϕ_n) of norm continuous and bounded functions on C. Lemma VI.6 gives that the functions $\tilde{\phi}_n(\mu) = \int \phi_n(x)\,d\mu(x)$ are norm continuous on $P(C)$. It follows by Lebesgues's theorem that the function $\tilde{\phi}(\mu) = \int \phi(x)\,d\mu(x)$ is norm lower semi-continuous on $P(C)$. Apply now Theorem II.11 to $\tilde{\phi}$ to obtain a function h in Lip(D) with $\|h\| \leq \varepsilon$ and such that $\mu \longrightarrow \int (\phi(x)+h(x))\,d\mu(x)$ attains its minimum on a Dirac measure δ_{x_0} for some x_0 in C. It is clear again from Theorem II.11 that h and x_0 may be chosen so that $\phi + h$ strongly exposes C from below at x_0.

2) => 1) As in the last part of proof of Theorem VI.1, one can use the *-Ekeland property (on $\phi(x) = d(x,K)$) to show that for each w*-compact subset

K of D such that $K \cap C = \emptyset$ there exists a w*-open neighborhood V such that $V \cap K \neq \emptyset$ and $d(C, \overline{V}^* \cap K) > 0$. By Lemma III.1, we can write K as a countable union of w*-compact sets that are a strictly positive distance away from C.

<u>Theorem VI.8</u>: Let X be a separable Banach space with the (P.C.P). Let C be any norm closed and bounded subset of X. Then for every bounded below lower semi-continuous function ϕ on C and every $\varepsilon > 0$, there exists a weakly continuous and norm-Lipschitz function h with $\|h\|_{\mathrm{Lip}(C)} \le \varepsilon$ and such that $\phi + h$ attains its minimum on C.

<u>Proof</u>: By Theorem IV.7 there exists a separable subspace Y of X^*, such that X is isometric to a closed subspace of Y^* in such a way that B_X is a strong w*-G_δ in B_{Y^*}. It follows from Lemma I.1 that any norm closed and bounded subset C of X is a strong w*-G_δ in Y^*, hence such sets have the *-Ekeland property by Theorem VI.7. Note now that the norm-Lipschitz perturbation h guaranteed by the *-Ekeland property for a ϕ and $\varepsilon > 0$ as in the statement of the theorem is $\sigma(C,Y)$ continuous hence obviously weakly continuous on C in X. The norm of h is computed in $\mathrm{Lip}(C)$.

We are now ready to give the following counterexamples:

<u>Example VI.9</u>: There exists a norm-closed, norm separable subset C of ℓ_∞ such that C is a w*-G_δ but not a strong w*-G_δ in ℓ_∞.

In particular, $P(C)$ is a separable closed convex and bounded w*-H_δ set but not a strong w*-H_δ in Lip*.

<u>Proof</u>: Denote by T the dyadic tree $\bigcup_{n=0}^{\infty} \{-1,1\}^n$ and let $\Gamma = \{-1,+1\}^{\mathbf{N}_*}$ where $\mathbf{N}_* = \mathbf{N} \setminus \{0\}$. Let $\Delta = \{\gamma \in \Gamma; \lim_n(\gamma_n) = -1\}$. We shall embed T and Γ in $\{-1,0,1\}^{\mathbf{N}}$ in the following way:

$$t = (\varepsilon_1, \varepsilon_2, \ldots, \varepsilon_n) \in T \longrightarrow (1, \varepsilon_1, \varepsilon_2, \ldots, \varepsilon_n, 0, 0, \ldots)$$

$$\gamma \in \Gamma \longrightarrow (1, \gamma_1, \gamma_2, \ldots \qquad)$$

We then equip $Z = T \cup \Gamma$ with the topology induced by $\{-1,0,1\}^{\mathbf{N}}$. Note that Z is then compact, Γ is closed in Z and the topology induced on Γ coincide with the topology of $\{-1,+1\}^{\mathbf{N}_*}$. Moreover T is countable, dense in Z and all its points are isolated in Z. Indeed, if $t^* = (\varepsilon_1^*,\ldots,\varepsilon_n^*) \in T$ and $V = \{s \in \{-1,0,1\}^{\mathbf{N}}; s_1 = 1,\ s_i = \varepsilon_i^* \text{ for } i = 1,\ldots,n,\ s_{n+1} = 0\}$ then V is open and $V \cap Z = \{t^*\}$.

Define now the following function ρ on T:

$$\rho(t) = \begin{cases} 1 \text{ if } t = \phi \quad \text{or } t = (\varepsilon_1,\ldots,\varepsilon_n), n \geq 1 \text{ and } \varepsilon_n = 1 \\ 2^{-k} \text{if } t = (\varepsilon_1,\ldots,\varepsilon_n),\ n \geq 1,\ \varepsilon_k = \varepsilon_{k+1} = \ldots = \varepsilon_n = -1 \\ \qquad\qquad \text{and } \varepsilon_{k-1} = 1 \text{ or if } k = 1. \end{cases}$$

For any $\gamma \in Z$ and positive integer n let $\gamma|n$ denote the element of T which agrees with γ in the first n places and is 0 elsewhere. For $t \in T$ let e_t denote the unit vector in $\ell_\infty(T)$ which is 1 at t and 0 elsewhere. Finally, for s and t in T, $s \leq t$ will mean that $s|n = t|n$ for $n = 1,\ldots,k$ for some positive integer k and that all coordinates of s beyond the kth one are 0.

Let now $\phi: Z \to \ell_\infty(T)$ be defined by:

$$\phi(t) = \sum_{s \leq t} \rho(s) e_s \quad \text{if} \quad t \in T$$

$$\phi(\gamma) = \sum_{n=0}^{\infty} \rho(\gamma|n) e_{(\gamma|n)} \quad \text{if} \quad \gamma \in \Gamma$$

Note that ϕ is one-to-one and is continuous from Z into (ℓ_∞, w^*) hence $D = \phi(Z)$ is homeomorphic to Z and is w^*-compact in ℓ_∞. Moreover, $C = \phi(T)$ is a w^*-G_δ (indeed it is relatively w^*-open in D), w^*-dense subset of D and $K = \phi(\Gamma)$ is w^*-compact. We shall show the following

(i) For each $\varepsilon > 0$, the set $K_\varepsilon = \{k \in K;\ d(k,C) \leq \varepsilon\}$ is w^*-dense in K.

(ii) The set $K^1 = \{k \in K; d(k,C) \geq 1\}$ is also w^*-dense in K.

In the terminology of Section III, this would mean that K is 1-bad with respect to C. In this case, K cannot be the countable union of w^*-compact subsets L_n of K such that $d(L_n,C) > 0$. Indeed if it were, then by Baire's

Category theorem one of the L_n's must have a non-empty interior relative to K which means that it intersects K_ε for each $\varepsilon > 0$; a contradiction.

To prove now (i) note that if $\gamma \in \Delta \setminus \Delta_n$ where $\Delta_n = \{\gamma; \gamma(j) = -1, \forall j \geq n\}$ we have $\|\phi(\gamma) - \phi(\gamma|k)\|_\infty \leq 2^{-n-1}$ for a large enough k. Hence K_ε is w*-dense in K for each $\varepsilon > 0$ since Δ is dense in Γ.

To prove (ii) note first that $C = D \cap c_0$ is norm closed and separable. Moreover, if $\gamma \notin \Delta$ then $\phi(\gamma)$ has infinitely many components equal to 1, hence $1 \leq d(\phi(\gamma), c_0) \leq d(\phi(\gamma), C)$ and $K \setminus \phi(\Delta) \subseteq K^1$ is w*-dense in K.

Before ending this section, we shall investigate the possibility of sharpening the above results in the following direction: Let D be a w^*-compact subset of a dual Banach space Y* and let C be a norm separable strong w*-G_δ subset of D. What are then the function spaces U such that for any bounded below and lower semi-continuous function ϕ on C, there exists arbitrarily small perturbation h in U so that $\phi + h$ strongly exposes C from below. Note that the above results show that the question is relevant only for spaces U such that $Y \subseteq U \subseteq \mathrm{Lip}(D)$. The optimal case is of course when Y = U; but this corresponds - in view of the results of Section II - to the case where C is a strong w*-H_δ set in D. The real question is then how better than Lip(D) can one do.

The methods of this section actually give the following

Proposition VI.10: Let U be a Banach space of functions such that $Y \subseteq U \subseteq \mathrm{Lip}(D)$ and which verifies the following properties:

(i) U is dense in C(D).

(ii) For any norm compact subset K in C, the function d(.,K) can be obtained as an iterated pointwise limit of a sequence of functions in the unit ball of U.

Then, for any bounded below lower semi-continuous function ϕ on C, the set $\{u \in U; \phi + u$ strongly exposes C from below$\}$ is a dense G_δ in U.

Proof: Note that (i) implies that on the set $\mathcal{P}(D)$ the $\sigma(\mathcal{P}(D), C(D))$ and $\sigma(\mathcal{P}(D), U)$ topologies coincide hence $\mathcal{P}(C)$ is a w^*-H_δ in $\mathcal{P}(D)$ considered as a subset of U^*.

Condition (ii) implies - as in the proof of Lemma VI.2 - that $\mathcal{P}(C)$ is norm separable in U^* and - as in the proof of Theorem VI.1 - that $\mathcal{P}(C)$ is a strong w^*-H_δ in $\mathcal{P}(D)$ considered as a subset of U^*. The rest of the proof is exactly as in Theorem VI.7.

To give a non-trivial example of such a space U, consider the vector subspace U_0 of Lip(D) consisting of the functions $u: Y^* \longrightarrow \mathbf{R}$ of the form $u = \phi \circ \Pi$ where $\Pi: Y^* \longrightarrow Y^*/N$ is a quotient map, N is a w^*-closed subspace of Y^* of finite codimension and ϕ is a bounded, Lipschitz and C^1-function from the finite dimensional space Y^*/N into $\mathbf{R}$. We denote by $\underline{\underline{C}}^1$ the closure of U_0 for the norm of Lip(D). It is easy then to check the following:

If $u \in \underline{\underline{C}}^1$ then:

(i) u is Frechet-differentiable everywhere on Y^* and for any y^* in Y^* the derivative $Du(y^*)$ of u at y^* is an element of Y.

(ii) The map $y^* \longrightarrow Du(y^*)$ from Y^* into Y is (w^*-to-$\|\ \|$) continuous.

In particular $\|u\|_{Lip(D)} = \max(\|Du\|; \sup_{y^* \in D}|u(y^*)|)$ where $\|Du\|$ is the norm of Du as an operator from Y^* into Y.

We shall need the following

Lemma VI.11: For any norm compact subset K of D, the function $d(.,K)$ is the iterated pointwise limit on D of functions in the unit ball of $\underline{\underline{C}}^1$.

Proof: Since K is norm compact, there exists a sequence of finite sets (F_n) such that $d(.,K)$ is the uniform limit of $d(.,F_n)$. Note now that there exists functionals (y_j) in the unit ball of Y such that for any

x^* in D we have

$$d(x^*,F_n) = \inf_{y^*\in F_n}\|x^*-y^*\| = \inf_{y^*\in F_n}\sup_j\langle x^*-y^*,y_j\rangle = \lim_m \uparrow \inf_{y^*\in F_n}\sup_{1\leq j\leq m}\langle x^*-y^*,y_j\rangle .$$

For each n,m, let $u_{nm}(x^*) = \inf_{y^*\in F_n}\sup_{1\leq j\leq m}\langle x^*-y^*,y_j\rangle$. It is clear that u_{nm} is a function in the unit ball of Lip(D) such that $d(x^*,K) = \lim_n \lim_m u_{nm}(x^*)$ for all x^* in D.

To show that u_{nm} can be chosen in $\underline{\underline{C}}^1$, it is enough to notice that a standard technique of regularization by convolution gives for any bounded and Lipschitz function u on $\mathbf{R}^N$ and any $\varepsilon > 0$, a function u' in $C^1(\mathbf{R}^N)$ such that $\|u-u'\|_\infty \leq \varepsilon$ and $\|u'\|_{\mathrm{Lip}(\mathbf{R}^N)} \leq \|u\|_{\mathrm{Lip}(\mathbf{R}^N)}$. This clearly finishes the proof of Lemma VI.11.

Now we can deduce the following

Theorem VI.12: Let D be a w^*-compact subset of a dual space Y^* and let C be a norm separable strong w^*-G_δ set in D. Then for any bounded below, norm lower semi-continuous function ϕ on C and any $\varepsilon>0$, there exists u in $\underline{\underline{C}}^1$ with $\|u\|_{C^1} \leq \varepsilon$ such that $\phi + u$ strongly exposes C from below.

Proof: Note that U_0 is a subalgebra of C(D) that contains the constant functions. Hence it is dense in C(D) by the Stone-Weierstrass theorem. This combined with Lemma VI.11 shows that $\underline{\underline{C}}^1$ verifies the hypothesis of Proposition VI.10, hence the conclusion of the theorem.

VII. APPLICATIONS TO NON-CONVEX OPTIMIZATION

In this section we show how the results established in the last one can be relevant in dealing with non-linear optimization problems. The reduction relies on a linearization procedure which consists of embedding a complete metric space in the dual of a Banach space in such a way that it becomes a strong w^*-G_δ set.

Let K be a space equipped with two bounded metrics Δ_1 and Δ_2. We shall say that the triplet (K,Δ_1,Δ_2) is a bi-metric space if Δ_2 is Δ_1-lower semi-continuous on $K \times K$.

A typical example of such a bitopological setting is the case of a bounded subset of the dual if a separable Banach space where Δ_1 is a metric that induces the w^*-topology and where Δ_2 is the norm. We shall show that this is actually the general case whenever (K,Δ_1) is assumed to be compact and (K,Δ_2) complete.

A subset C of K will be called a Δ_2-strong Δ_1-G_δ in K if $K \setminus C = \bigcup_n F_n$ where each F_n is Δ_1-closed and Δ_2-$d(F_n,C) > 0$ for each n. Again, the typical example is the case of a strong w^*-G_δ set in some dual Banach space. The following Theorem shows that this is actually the general case.

Theorem VII.1: Let (K,Δ_1,Δ_2) be a bi-metric space such that (K,Δ_1) is compact and (K,Δ_2) is complete. There exists then a Banach space Y and a map $\delta : K \longrightarrow Y^*$ such that

(i) δ is a homeomorphism between (K,Δ_1) and $(\delta(K),w^*)$.

(ii) There exists $\alpha > 0$ such that for any (x,y) in $K \times K$ we have

$$\|\delta_x - \delta_y\| \leq \Delta_2(x,y) \leq \alpha\|\delta_x - \delta_y\|.$$

The space Y is actually the space of Δ_2-Lipschitz and Δ_1-continuous functions on K equipped with the norm $\|f\| = \|f\|_{Lip(\Delta_2)} + \|f\|_\infty$.

For the proof, we need the following

Lemma VII.2: Let (K,Δ_1,Δ_2) be a bi-metric space such that (K,Δ_1) is compact, then for any Δ_1-compact sets L_0 and L_1 in K such that $\Delta_2\text{-dist}(L_0,L_1) > 1$, there exists a Δ_1-continuous and Δ_2-Lipschitz function ϕ on K such that:

1) $0 \leq \phi \leq 1$

2) $\|\phi\|_{Lip(\Delta_2)} \leq 4$

3) $\phi = 0$ on L_0 and $\phi = 1$ on L_1.

Proof: For any subset A of K, we denote by $B_2(A,\varepsilon)$ the set $\{x \in K;\ \Delta_2(x,A) \leq \varepsilon\}$. Note that if A is Δ_1-closed then $B_2(A,\varepsilon)$ is also Δ_1-closed since Δ_2 is Δ_1-l.s.c. Let $U_0 = L_0$, $U_1 = L_1^c$; $V_i = U_i^c$ for $i = 0,1$. We shall construct, by induction, for each dyadic number θ in $]0,1[$, two Δ_1-open sets U_θ and V_θ such that

a) $U_\theta \cap V_\theta = \emptyset$

b) $U_{\frac{2k+1}{2^{\ell+1}}} \supseteq B_2(V^c_{k/2^\ell}, 2^{-\ell-1})$ for each ℓ and $k=0,1,\dots,2^\ell$

c) $V_{\frac{2k+1}{2^{\ell+1}}} \supseteq B_2(U^c_{k+1/2^\ell}, 2^{-\ell-1})$ for each ℓ and $k=0,1,\dots,2^\ell$.

For that let $F_0^{(1)} = B_2(L_0,1/2)$ and $G_1^{(1)} = B_2(L_1,1/2)$. Since $\Delta_2\text{-}d(L_0,L_1) > 1$, they are 2 disjoint Δ_1-closed sets, hence there exists two disjoint Δ_1-open sets $U_{1/2}$ and $V_{1/2}$ such that $F_0^{(1)} \subseteq U_{1/2}$ and $G_1^{(1)} \subseteq V_{1/2}$. By construction we have that $U_{1/2} \supseteq B_2(V_0^c, 1/2)$ and $V_{1/2} \supseteq B_2(U_1^c, 1/2)$. Suppose now (U_θ, V_θ) constructed for all θ in $\{k/2^n;\ k=0,1,\dots,2^n\} = \Delta_n$. For each θ in Δ_n, set $F_\theta^{(n+1)} = B_2(V_\theta^c, 2^{-n-1})$ and $G_\theta^{(n+1)} = B_2(U_\theta^c, 2^{-n-1})$. We claim that for each k in $\{0,1,\ldots,2^n-1\}$, $F^{(n+1)}_{k/2^n}$ and $G^{(n+1)}_{k+1/2^n}$ are disjoint. Indeed, assume $k=2\ell$. (The odd case will be similar) we have from b) that

(*) $U_{k+1/2^n} = U_{2\ell+1/2^n} \supseteq B_2(V^c_{2\ell/2^n}, 2^{-n})$.

If there exists f such that $\Delta_2(f,x) \le 2^{-n-1}$ and $\Delta_2(f,y) \le 2^{-n-1}$ for some x in $V^c_{k/2^n}$ and y in $U^c_{k+1/2^n}$ then $\Delta_2(x,y) \le 2^{-n}$ and y would belong to $B_2(V^c_{k/2^n}, 2^{-n})$ which contradicts (*).

Separate now $F^{(n+1)}_{k/2^n}$ and $G^{(n+1)}_{k+1/2^n}$ by two disjoint Δ_1-open sets $U_\theta = U_{2k+1/2^{n+1}}$ and $V_\theta = V_{2k+1/2^{n+1}}$. It is easily verified that (U_θ, V_θ) satisfy the hypothesis a) b) c).

Note now that if θ and θ' are two dyadic numbers such that $\theta < \theta'$ then $K = U_{\theta'} \cup V_\theta$. It follows that the function defined by $\phi(x) = \sup\{\theta \in [0,1];\ x \in V_\theta\}$ is Δ_1-continuous. Moreover $\phi = 0$ on L_0 and $\phi = 1$ on L_1. To show that ϕ is Δ_2-Lipschitz, let x and y be in K such that $\phi(x) = s$ and $\phi(y) = t$ with $s < t$. We can find two dyadic numbers $k/2^n$ and $k+1/2^n$ such that $s < k/2^n < k+1/2^n < t$ and $(t-s) \le 4.2^{-n}$. Hence $x \in U_{k/2^n}$ and $y \in V_{k+1/2^n}$. If k is odd we have

$$V_{k/2^n} \supseteq B_2(U^c_{k+1/2^n}, 2^{-n}) \supseteq B_2(V_{k+1/2^n}, 2^{-n}).$$

Hence $x \in U_{k/2^n} \Rightarrow x \notin V_{k/2^n} \Rightarrow \Delta_2(x, V_{k+1/2^n}) > 2^{-n}$ and $\Delta_2(x,y) > 2^{-n}$. It follows that $|\phi(x) - \phi(y)| = (t-s) \le 4\ 2^{-n} \le 4.\ \Delta_2(x,y)$. The case where k is even is similar. This finishes the proof of Lemma VII.2.

To prove Theorem VII.1, assume (K,Δ_1) compact and (K,Δ_2) complete. Let $C(K,\Delta_1)$ be the Banach space of Δ_1-continuous functions on K. Let Y be the Banach space of all Δ_1-continuous and Δ_2-Lipschitz functions on K equipped with the norm $\|f\| = \|f\|_{Lip(\Delta_2)} + \|f\|_\infty$. Note that Y is a vector sublattice of $C(K,\Delta_1)$ that separates the points of K by Lemma VII.2. Hence Y is norm dense in $C(K,\Delta_1)$ by the Stone-Weierstrass theorem. Let now $i:Y \longrightarrow C(K,\Delta_1)$ be the canonical injection, we get that the adjoint map i* is an homeomorphism for the respective w*-topologies on the w*-compact subsets of $C(K,\Delta_1)^*$. Embed now K in Y* by associating to each x in K the functional $\delta_x(\phi) = \phi(x)$ for each ϕ in Y. It follows that the set $\{\delta_x; x \in K\}$ is the image by i* of the set of Dirac measures in $C(K,\Delta_1)^*$. Hence (K,Δ_1) is homeomorphic to $\{\delta_x; x \in K\}$ equipped with the $\sigma(Y^*,Y)$ topology. On the other hand for each x and y in K we have:

$$\|\delta_x - \delta_y\|_{Y^*} = \sup_{\substack{f \in Y \\ \|f\| \le 1}} |f(x) - f(y)| \le \Delta_2(x,y).$$

Moreover, by Lemma VII.2, we get for each $0 < \varepsilon < 1$, a function f in Y with $\|f\|_{Lip(\Delta_2)} \le 4$, $0 \le f \le (1-\varepsilon)\Delta_2(x,y)$ such that $f(x) = 0$ and $f(y) = (1-\varepsilon)\Delta_2(x,y)$. Hence

$$\Delta_2(x,y) = \frac{1}{1-\varepsilon} |f(x) - f(y)| \le \frac{1}{1-\varepsilon} \|f\| \, \|\delta_x - \delta_y\| \le \left(\frac{4}{1-\varepsilon} + M\right) \|\delta_x - \delta_y\|$$

where M is any upper bound for Δ_2 on $K \times K$.

Q.E.D.

Suppose now that C is a Δ_2-separable, Δ_1-G_δ in K. The Baire category theorem gives that C contains a Δ_1-dense set of points of $(\Delta_1$-$\Delta_2)$ continuity. Let us call a point x in C a <u>$(\Delta_1$-$\Delta_2)$ peak point</u> if there exists a Δ_1-continuous and Δ_2-Lipschitz function h on C such that

(i) h attains its maximum on C at x

(ii) Every maximizing sequence for h, Δ_2-converges to x.

Note that the $(\Delta_1 - \Delta_2)$ peak points are necessarily points of $(\Delta_1$-$\Delta_2)$ continuity.

By combining Theorem VII.1, Theorem VI.7 and proposition VI.5 we obtain the following extension of a Theorem of Ekeland [12]. See also Corollary VII.5.

Theorem VII.3: Let (K,Δ_1,Δ_2) be a bi-metric space such that (K,Δ_1) is compact and (K,Δ_2) is complete. Let C be a Δ_2-separable, Δ_2-strong Δ_1- G_δ in K. Then:

1) C contains a Δ_1-dense set of $(\Delta_1-\Delta_2)$ peak points.

2) For any bounded below Δ_2-lower semi-continuous function ϕ on C and any $\varepsilon > 0$, there exists a Δ_1-continuous and Δ_2-Lipschitz function h with $\|h\|_{Lip(\Delta_2)} \leq \varepsilon$ and an x_0 in C such that

(i) $(\phi+h)$ attains its minimum on C at x_0.

(ii) Every minimizing sequence for $\phi + h$ must Δ_2-converge to x_0.

Proof: It is enough to notice that C can be identified with a strong w*-G_δ set in the dual of the space Y considered in Theorem VII.1.

To show that Ekeland's theorem is a special case of the above we need the following.

Proposition VII.4: Let (C,d) be a separable complete metric space such that d is bounded on C. Then there exists a bi-metric space (K,Δ_1,Δ_2) such that (K,Δ_1) is compact, (K,Δ_2) is complete, C is a Δ_2-strong Δ_1-G_δ in K and $\Delta_2 = d$ on C.

Proof: We may and shall assume without loss of generality that the metric d is bounded by one on C. Let Lip(C) denote the Banach space of all bounded and d-Lipschitz functions on C equipped with the norm $\|f\|=\|f\|_\infty + \|f\|_L$. We embed C in Lip (C)* by associating to each point x in C the functional $\delta_x(\phi) = \phi(x)$ for each $\phi \in$ Lip (C). We compactify C by taking $K = \overline{\{\delta_x; x \in C\}}^*$ in Lip (C)*. Note that $d(x,y) = \|\delta_x - \delta_y\|_{Lip\ (C)^*}$ for all x,y in C and the norm can be viewed as an extension of the metric d to K. We claim that the set K is w*-metrizable in Lip (C)*. Indeed, if T is a countable dense subset of C, we consider the subspace Y of Lip (C) generated by the constant one and the finite supremums of the functions $\{\phi_x; x \in T\}$ where ϕ_x is defined by $\phi_x(y) = d(y,x)$. Note that Y is separable and is dense in Lip(C) for the norm of $\ell_\infty(C)$. This implies that the topologies $\sigma(K,\mathrm{Lip}(C))$ and $\sigma(K,Y)$ coincide since the elements of K are bounded on the functions of Lip(C) once equipped with the $\ell_\infty(C)$-norm.

Claim (1): For each $x \in C$, the function $\mu \to \|\mu - \delta_x\|$ is w*-continuous on K.

Proof: Suppose (y_n) is a sequence in C such that $\mu = \text{w*-}\lim(\delta_{y_n})$ in Lip (C)*. Note that $\phi_x = d(\cdot,x)$ belongs to the ball of Lip(C), hence

$$\mu(\phi_x) = \lim_n \phi_x(y_n) = \lim_n d(y_n,x) \quad \text{and}$$

$$\|\mu - \delta_x\| \geq \langle \mu - \delta_x, \phi_x\rangle = \langle \mu, \phi_x\rangle = \lim_n d(y_n,x).$$

On the other hand if ϕ belongs to the ball of Lip (C) we have

$$|\langle \mu - \delta_x, \phi\rangle| = \lim_n |\phi(y_n) - \phi(x)| \leq \lim_n d(y_n,x).$$

It follows that if $\delta_{y_n} \xrightarrow{w^*} \mu$ then $\|\delta_{y_n} - \delta_x\| = d(y_n,x)$ converges to $\|\mu - \delta_x\|$ which clearly proves the claim.

Claim (2): The set C is a strong w*-G_δ in K.

Proof: Since for each x in C, the functions $\mu \to \|\mu - \delta_x\|$ are w*-continuous on K, we get that the metric d induces on C the w*-topology and since (C,d) is complete metric, it follows that C is a w*-G_δ in K.

To show that C is a strong w*-G_δ subset of K, note that the function $\mu \to d(\mu,C) = \inf\{\|\mu - \delta_x\|; x \in C\}$ is weak* upper-semi-continuous on K. Hence if L is any w*-compact subset of K such that $L \cap C = \phi$ the set $L_\varepsilon = \{\ell \in L; d(\ell,C) < \varepsilon\}$ is w*-open for each $\varepsilon > 0$. We claim that L_ε cannot be w*-dense in L for each $\varepsilon > 0$. Indeed, if not then by Baire's category theorem the set $\bigcap_{\varepsilon>0} L_\varepsilon$ will be non-empty and will intersect C since the latter is norm-closed; a contradiction. Hence there exists $\varepsilon > 0$ such that $L\setminus \overline{L}^*_\varepsilon \neq \phi$. Take any x in $L\setminus \overline{L}^*_\varepsilon$ and any w*-neighborhood of x such that $V \cap \overline{L}^*_\varepsilon = \phi$. Note then that $V \cap L \neq \phi$ and $d(V \cap L, C) > 0$. Use now Lemma III.1 to show that L can be written as a countable union of w*-compact sets (L_n) such that $d(L_n, C) > 0$.

To finish the proof of Claim (2), write $K\setminus C = \bigcup_n K_n$ where each K_n is a w*-compact set. Apply the above remark to split each K_n into a countable union of w*-compact sets $K_{n,m}$ such that $d(K_{n,m}, C) > 0$. Hence C is a strong w*-G_δ set in K.

If now Δ_1 is any metric that induces the w*-topology on K, and if Δ_2 is the metric induced by the norm then the triplet (K,Δ_1,Δ_2) clearly verifies the claim of Proposition VII.3.

Now we can deduce Ekeland's theorem.

<u>Corollary VII.5</u>: Let (C,d) be a separable complete metric space and let ϕ be a bounded below lower semi-continuous function on C. Then for each $\varepsilon > 0$, there exists x_0 in C such that: $\phi(x) \geq \phi(x_0) - \varepsilon d(x,x_0)$ for all x in C.

<u>Proof</u>: We can assume without loss that d is bounded by 1. If not we simply take $\bar{d} = \min(d,1)$. Let (K,Δ_1,Δ_2) be the triplet associated to (C,d) by the above proposition. Apply Theorem VII.3 to get a Δ_2-Lipschitz, Δ_1-continuous function h with $\|h\|_{Lip(\Delta_2)} \leq \varepsilon$ and x_0 in C such that:

$$\phi(x) \geq \phi(x_0) - (h(x) - h(x_0)) \geq \phi(x_0) - \varepsilon\,\Delta_2(x,x_0).$$

Finally note that $\Delta_2 = d$ on C.

We used several times in this paper the fact that for any G_δ-subset C of a metric space (K,Δ_1) there exists a metric Δ on K that induces the Δ_1-topology on C while making it complete. For strong G_δ-sets we have the following.

Theorem VII.6: Let (K,Δ_1,Δ_2) be a bi-metric space such that (K,Δ_1) is compact. Suppose $\Delta_2 \geq \Delta_1$ and let C be a Δ_2-strong, Δ_1-G_δ in K. Then there exists a metric Δ on K that verifies the following properties:

1) $\Delta \leq \Delta_2$ on $K \times K$.

2) Δ is Δ_1-lower semi-continuous on $K \times K$ and induces the Δ_1-topology on C.

3) (C,Δ) is complete.

Lemma VII.7: Let (K,Δ_1,Δ_2) be a bi-metric space such that (K,Δ_1) is compact and $\Delta_2 \geq \Delta_1$. Let L be a Δ_1-compact subset and let Ω_ρ, be the Δ_1-open set $\{x \in K; \Delta_2(x,L) > \rho\}$ where $\rho > 0$. There exists then a metric Δ_ρ verifying:

(1) $\Delta_\rho \leq \Delta_2$.

(2) Δ_ρ is Δ_1-lower semi-continuous on $K \times K$ and induces the Δ_1-topology on Ω_ρ.

(3) No point of L is a Δ_1-limit of a Δ_ρ-Cauchy sequence in Ω_ρ.

Proof: Let (U_n) be a sequence of Δ_1-open subsets of Ω_ρ such that $U_n \subseteq \overline{U}_n \subseteq U_{n+1}$ for all n and $\Omega_\rho = \bigcup_n U_n$. (The closure being taken in Δ_1). Since $\Delta_2(\overline{U}_n,L) > \rho$ for each n, apply Lemma VII.2 to each pair $(\overline{U}_n,K)$ to obtain a Δ_1-continuous function ϕ_n with $\|\phi_n\|_{Lip(\Delta_2)} \leq 4$, $\phi_n = \rho$ on $\overline{U}_n$, $\phi_n = 0$ on L and $0 \leq \phi_n \leq \rho$. Let now

$$\Delta_\rho(x,y) = \Delta_1(x,y) + \sup_n |\phi_n(x) - \phi_n(y)|.$$

Note first that $\Delta_\rho \leq 5\,\Delta_2$. Obviously Δ_ρ is Δ_1-lower semi-continuous. On the other hand, if ω is a Δ_1-open set contained in some U_m we have for any x and y in ω that $\phi_n(x) = \phi_n(y) = \rho$ for all $n \geq m$, hence $\Delta_\rho(x,y) = \Delta_1(x,y) + \sup_{1 \leq n \leq m} |\phi_n(x) - \phi_n(y)|$ is Δ_1-continuous on ω. It follows that Δ_ρ induces the Δ_1-topology on Ω_ρ.

The function $\phi_\infty = \sup_n \phi_n$ is Δ_ρ-continuous and is equal to ρ on Ω_ρ, hence $\phi_\infty = \rho$ on the Δ_ρ-closure of Ω_ρ. On the other hand $\phi_\infty = 0$ on L. If now (x_ℓ) is a Δ_ρ-Cauchy sequence in Ω_ρ that Δ_1-converges to a point x in L, we get since Δ_ρ is Δ_1-lower semi-continuous that $\lim_\ell \Delta_\rho(x,x_\ell) = 0$, hence x is in the Δ-closure of Ω_ρ and in L, a contradiction.

To prove Theorem VII.6, write $K \setminus C = \bigcup_n L_n$ where each L_n is Δ_1-compact such that $\Delta_2(L_n,C) > \rho_n > 0$. Let Δ_n be the metric constructed above for $\Omega_n = \{x \in K;\ \Delta_2(x,L_n) > \rho_n\}$. It is easy to see that the metric $\Delta = \sum_n 2^{-n}\Delta_n$ verifies the claims of the theorem.

An immediate corollary of Theorem VII.6 is the following.

Corollary VII.8: Let C be a bounded subset of a dual Banach space Y* such that $K = \overline{C}^*$ is w*-metrizable. If C is a strong w*-G_δ in K then there exists a norm-Lipschitz and w*-lower semi-continuous metric Δ on K that induces the w*-topology on C while making it complete.

REFERENCES

[1] J. Bourgain: Sets with the Radon-Nikodym property in conjugate Banach spaces. Studia Math. T.LXVI (1978) p.199-205.

[2] J. Bourgain: La propriété de Radon-Nikodym. Publications Mathématiques de l'Université Pierre et Marie Curie - No.36 (1979).

[3] J. Bourgain, H. P Rosenthal: Geometrical implications of certain finite dimensional decompositions. Bull. Soc. Math. Belg. 32, (1980) p.57-82.

[4] R. D. Bourgin: Geometric Aspects of convex sets with the Radon-Nikodym property. Springer-Verlag No. 993 (1983).

[5] A. Brondsted, R. T. Rockafellar: On the subdifferentiability of convex functions. Proc. A.M.S. 16 (1965) p. 605-611.

[6] G. Choquet: Lectures in Analysis. Vol. II. W. A. Benjamin, Inc., New-York (1969).

[7] W. Davis, N. Ghoussoub, J. Lindenstrauss: A lattice renorming theorem and applications to vector-valued processes. T.A.M.S. 263 p.531-540 (1981).

[8] W. Davis, W. B. Johnson: A renorming of non-reflexive Banach spaces. Proc. A.M.S. 37 p.486-487 (1973).

[9] J. Diestel: Geometry of Banach Spaces - Selected topics. Lecture notes in Math. Vol. 485 Springer-Verlag (1975).

[10] J. Diestel, J. Jr. Uhl: Vector measures. Math Surveys, 15 A.M.S. (1977).

[11] G. A. Edgar, R. F. Wheeler: Topological properties of Banach spaces. Pac. J. Math. 115, p.317-350 (1984).

[12] I. Ekeland: Non convex minimization problems. Bull. A.M.S. 1 (1979) p. 443-474.

[13] D. H. Fremlin, M. Talagrand: On CS closed sets. Mathematica, 26 (1979) p. 30-32.

[14] N. Ghoussoub, B. Maurey: G_δ-embeddings in Hilbert space. J. Funct. Analysis (1984) Vol.61, No. 1, p.72-97 (1985).

[15] N. Ghoussoub, B. Maurey: The Asymptotic norming and the Radon-Nikodym properties are equivalent in separable Banach spaces. Proc. A.M.S. Vol. 94, 4, p.665-671 (1985).

[16] N. Ghoussoub, B. Maurey: The Radon-Nikodym property in function spaces. Proceedings of C.B.M.S. conference at Missouri (1984). To appear.

[17] J. E. Jayne, C. A. Rogers: The extremal structure of convex sets. J. Funct. Analysis (3) 26, p. 251-288 (1977).

[18] R. C. James, A. Ho: The asymptotic norming and Radon-Nikodym properties for Banach spaces. Arkiv for Matematik 19, p.53-70 (1981).

[19] J. L. Kelley: General Topology. Van Nostrand Company. (1955).

[20] Y. Lindenstrauss, L. Tzafriri: Classical Banach spaces II-Function spaces. Springer-Verlag 97 (1979).

[21] H. P. Lotz: Extensions and liftings of positive linear operators T.A.M.S. 211, p.85-100 (1975).

[22] J. Neveu: Discrete parameter martingales. North Holland (1975).

[23] E. Odell, H. P. Rosenthal: A double-dual characterization of separable Banach spaces containing ℓ_1. Israel J. Math 20 (1975) p. 375-384.

[24] R. R. Phelps: Dentability and extreme points in Banach spaces. J. Funct. Analysis (1) 17 (1974) p. 78-90.

[25] H. P. Rosenthal: Geometric properties related to the Radon-Nikodym property. Seminaire d'Initiation a l'analyse, 20^{e} Annêe, N° 22, (1980-1981).

[26] C. Stegall: The Radon-Nikodym property in conjugate Banach spaces I. T.A.M.S. 206 (1975) p.213-223.

[27] C. Stegall: The Radon-Nikodym property in conjugate Banach spaces II. T.A.M.S. 264 (1981) p. 507-519.

[28] C. Stegall: Optimization of functions on certain subsets of Banach spaces. Math Annalen, 236 (1978) p.171-176.

[29] M. Talagrand: La structure des espaces de Banach rêticulês ayant la proprietê de Radon-Nikodym. Israel J. Math. 44 No.3 (1983).

[30] H. V. Weiszacker: A note on infinite dimensional convex sets. Math Scand. 38 p.321-324 (1976).

[32] D. Dacunha-Castelle, M. Schreiber: Techniques probabilistes pour l'étude des problemes d'isomorphismes entre espaces de Banach. Ann. Inst. H. Poincaré 10, p.229-277 (1974).

NASSIF GHOUSSOUB. The University of British Columbia, Vancouver, Canada.

BERNARD MAUREY. Universite Paris VII, Paris France.

General instructions to authors for
PREPARING REPRODUCTION COPY FOR MEMOIRS

For more detailed instructions send for AMS booklet, "A Guide for Authors of Memoirs." Write to Editorial Offices, American Mathematical Society, P. O. Box 6248, Providence, R. I. 02940.

MEMOIRS are printed by photo-offset from camera copy fully prepared by the author. This means that, except for a reduction in size of 20 to 30%, the finished book will look exactly like the copy submitted. Thus the author will want to use a good quality typewriter with a new, medium-inked black ribbon, and submit clean copy on the appropriate model paper.

Model Paper, provided at no cost by the AMS, is paper marked with blue lines that confine the copy to the appropriate size. Author should specify, when ordering, whether typewriter to be used has PICA-size (10 characters to the inch) or ELITE-size type (12 characters to the inch).

Line Spacing – For best appearance, and economy, a typewriter equipped with a half-space ratchet – 12 notches to the inch – should be used. (This may be purchased and attached at small cost.) Three notches make the desired spacing, which is equivalent to 1-1/2 ordinary single spaces. Where copy has a great many subscripts and superscripts, however, double spacing should be used.

Special Characters may be filled in carefully freehand, using dense black ink, or INSTANT ("rub-on") LETTERING may be used. AMS has a sheet of several hundred most-used symbols and letters which may be purchased for $5.

Diagrams may be drawn in black ink either directly on the model sheet, or on a separate sheet and pasted with rubber cement into spaces left for them in the text. Ballpoint pen is *not* acceptable.

Page Headings (Running Heads) should be centered, in CAPITAL LETTERS (preferably), at the top of the page – just above the blue line and touching it.

LEFT-hand, EVEN-numbered pages should be headed with the AUTHOR'S NAME;

RIGHT-hand, ODD-numbered pages should be headed with the TITLE of the paper (in shortened form if necessary).

Exceptions: PAGE 1 and any other page that carries a display title require NO RUNNING HEADS.

Page Numbers should be at the top of the page, on the same line with the running heads.

LEFT-hand, EVEN numbers – flush with left margin;

RIGHT-hand, ODD numbers – flush with right margin.

Exceptions: PAGE 1 and any other page that carries a display title should have page number, centered below the text, on blue line provided.

FRONT MATTER PAGES should be numbered with Roman numerals (lower case), positioned below text in same manner as described above.

MEMOIRS FORMAT

It is suggested that the material be arranged in pages as indicated below.
Note: <u>Starred items (*)</u> <u>are requirements of publication.</u>

Front Matter (first pages in book, preceding main body of text).

Page i – *Title, *Author's name.

Page iii – Table of contents.

Page iv – *Abstract (at least 1 sentence and at most 300 words).

*<u>*1980 Mathematics Subject Classification (1985 Revision).*</u> This classification represents the primary and secondary subjects of the paper, and the scheme can be found in Annual Subject Indexes of MATHEMATICAL REVIEWS beginning in 1984.

Key words and phrases, if desired. (A list which covers the content of the paper adequately enough to be useful for an information retrieval system.)

Page v, etc. – Preface, introduction, or any other matter not belonging in body of text.

Page 1 – Chapter Title (dropped 1 inch from top line, and centered).

Beginning of Text.

Footnotes: *Received by the editor date.

Support information – grants, credits, etc.

Last Page (at bottom) – Author's affiliation.

ABCDEFGHIJ – 89876